Soft Computing in Textile Sciences

Studies in Fuzziness and Soft Computing

Further volumes of this series can be found at our homepage.

Vol. 88. R.P. Srivastava and T.J. Mock (Eds.)
Belief Functions in Business Decisions, 2002
ISBN 3-7908-1451-2

Vol. 89. B. Bouchon-Meunier, J. Gutiérrez-Ríos, L. Magdalena and R.R. Yager (Eds.)
Technologies for Constructing Intelligent Systems 1, 2002
ISBN 3-7908-1454-7

Vol. 90. B. Bouchon-Meunier, J. Gutiérrez-Ríos, L. Magdalena and R.R. Yager (Eds.)
Technologies for Constructing Intelligent Systems 2, 2002
ISBN 3-7908-1455-5

Vol. 91. J.J. Buckley, E. Eslami and T. Feuring
Fuzzy Mathematics in Economics and Engineering, 2002
ISBN 3-7908-1456-3

Vol. 92. P.P. Angelov
Evolving Rule-Based Models, 2002
ISBN 3-7908-1457-1

Vol. 93. V.V. Cross and T.A. Sudkamp
Similarity and Compatibility in Fuzzy Set Theory, 2002
ISBN 3-7908-1458-X

Vol. 94. M. MacCrimmon and P. Tillers (Eds.)
The Dynamics of Judicial Proof, 2002
ISBN 3-7908-1459-8

Vol. 95. T.Y. Lin, Y.Y. Yao and L.A. Zadeh (Eds.)
Data Mining, Rough Sets and Granular Computing, 2002
ISBN 3-7908-1461-X

Vol. 96. M. Schmitt, H.-N. Teodorescu, A. Jain, A. Jain, S. Jain and L.C. Jain (Eds.)
Computational Intelligence Processing in Medical Diagnosis, 2002
ISBN 3-7908-1463-6

Vol. 97. T. Calvo, G. Mayor and R. Mesiar (Eds.)
Aggregation Operators, 2002
ISBN 3-7908-1468-7

Vol. 98. L.C. Jain, Z. Chen and N. Ichalkaranje (Eds.)
Intelligent Agents and Their Applications, 2002
ISBN 3-7908-1469-5

Vol. 99. C. Huang and Y. Shi
Towards Efficient Fuzzy Information Processing, 2002
ISBN 3-7908-1475-X

Vol. 100. S.-H. Chen (Ed.)
Evolutionary Computation in Economics and Finance, 2002
ISBN 3-7908-1476-8

Vol. 101. S.J. Ovaska and L.M. Sztandera (Eds.)
Soft Computing in Industrial Electronics, 2002
ISBN 3-7908-1477-6

Vol. 102. B. Liu
Theory and Practice of Uncertain Programming, 2002
ISBN 3-7908-1490-3

Vol. 103. N. Barnes and Z.-Q. Liu
Knowledge-Based Vision-Guided Robots, 2002
ISBN 3-7908-1494-6

Vol. 104. F. Rothlauf
Representations for Genetic and Evolutionary Algorithms, 2002
ISBN 3-7908-1496-2

Vol. 105. J. Segovia, P.S. Szczepaniak and M. Niedzwiedzinski (Eds.)
E-Commerce and Intelligent Methods, 2002
ISBN 3-7908-1499-7

Vol. 106. P. Matsakis and L.M. Sztandera (Eds.)
Applying Soft Computing in Defining Spatial Relations, 2002
ISBN 3-7908-1504-7

Vol. 107. V. Dimitrov and B. Hodge
Social Fuzziology, 2002
ISBN 3-7908-1506-3

Les M. Sztandera
Christopher Pastore
Editors

Soft Computing in Textile Sciences

With 70 Figures
and 22 Tables

Springer-Verlag Berlin Heidelberg GmbH
A Springer-Verlag Company

Professor Dr. Les M. Sztandera
Philadelphia University
School of Business Administration
Computer Information Systems
Philadelphia, PA 19144
USA
sztanderal@philau.edu

Professor Dr. Christopher Pastore
Philadelphia University
Textile Engineering Department
Philadelphia, PA 19144
USA
pastorec@philau.edu

DOI 10.1007/978-3-7908-1750-8

Library of Congress Cataloging-in-Publication Data applied for
Die Deutsche Bibliothek – CIP-Einheitsaufnahme
Soft computing in textile sciences: with 22 tables / Les M. Sztandera; Christopher Pastore ed. – Heidelberg; New York: Physica-Verl., 2003
(Studies in fuzziness and soft computing; Vol. 108)

Originally published by Physica-Verlag Heidelberg in 2003
MyCopy version of the original edition 2003

Printed on acid-free paper
www.springer.com/mycopy

Preface

Textiles and computing have long been associated. A century ago the prototype for the punch card was developed by textile machinery producers. One of the earliest commercial software algebraic interpreters (originally Q&A, later TKSolver) was developed by a textile engineer, Milos Konopasek, to solve problems in the textile field. We can find more examples of this kind, but suffice it to say that the high volume and low profit margin of textile products has driven the industry to invest in high technology, particularly in the area of data interpretation and analysis. Thus it is virtually inevitable that soft computing has found a home in the textile industry.

The early roots of soft computing could be traced back to Dr. Lotfi A. Zadeh's book chapter on *soft data analysis* [1] published in 1981. Nevertheless, the actual concept of 'soft computing' (SC) was launched about 10 years later, when the Berkeley Initiative in Soft Computing (BISC), an industrial liaison program, was established at the University of California – Berkeley.

The main characteristics of soft computing as relevant to this work are:

- Capability to approximate various kinds of real-world systems;
- Tolerance for imprecision, partial truth, and uncertainty; and
- Ability to learn from the environment.

These characteristics are commonly leading to better model of reality, low solution cost, robustness, and tractability. Dr. Zadeh, the father of soft computing, has emphasized that soft computing provides a solid foundation for the conception, design, and application of intelligent systems employing its member methodologies *symbiotically* rather than in isolation.

The soft computing community is not looking for perfect solutions but rather competitive ones. Soft computing can serve a critical role in provide temporary solutions to problems wherein the fundamental theoretical background has not been completed, or the complexity of the problem escapes first principle type solutions. There exists an implicit commitment to benefit from the fusion of various methodologies. Such a fusion can lead to cooperative and complementary combinations of the individual methodologies. Constructive fusion thinking has already been extended beyond the individual SC technologies.

The intent of this book is to put together under one cover original contributions by authors who have made significant contributions in both the field of soft computing and the field of textiles and apparel. There are six chapters in this edited volume discussing various aspects of soft computing in the field of textile science.

The textile industry is rather broad – ranging from the growth of plants and animals (for production of cotton and wool) to synthetic chemical processing (for fibers such as nylon, polyester, *etc.*), to rapid processing for converting fibers to yarns and yarns to fabrics. The industry also includes significant amounts of chemical processes for dyeing and finishing fabrics, computer controlled cutting devices for making clothing, and then gets into business logistics of inventory, transportation, marketing, management, and more. This broad field identified as "textiles" is represented in this book. This includes textiles and apparel.

Chapter 1, by Fang *et al.*, presents a discussion of the soft goods supply chain using fuzzy logic and neural networks to model the integration of the suppliers and consumers. Soft goods refers to the entire sequence from fiber to finished product. The model addresses capacity allocation and delivery date assignment. A simulation of a typical supply chain is presented which is realized through the application of soft computing techniques.

Composite mechanics is the topic of Chapter 2, wherein Muc and Kedziora apply principles of fuzzy logic to the solution of a challenging mechanics problem. This is an exciting application of soft computing to a classical mechanics problem of failure of composites. The classical problem falls short in that manufacturing of composite materials inevitably involves manufacturing variability. Accounting for these artifacts has not been achieved through first principles, and the use of soft computing techniques is shown to be successful in this work.

Sztandera *et al.* develop a soft computing approach in the third Chapter that employ neural networks and fuzzy logic to model the potential toxicity of as yet undeveloped dye chemicals. The method makes use of a deep understanding of molecular chemistry with a robust database on carcinogenecity of various dye chemicals. Again we observe the use of soft computing with a physical problem wherein the fundamental principles are not yet known. A new concept of fuzzy entropy is presented in this chapter, making it an exciting contribution to the book.

In Chapter 4, Xu presents a solution to the difficult problem of color matching. The human eye is exceptionally sensitive to slight variations in color, and the need to match color exactly and repeatedly throughout production is critical to successful product development. The use of fuzzy logic and neural networks is applied to solve this important challenge.

The fifth chapter of this book presents the work of Brannon *et al.* wherein they attempt to model the retail environment using agent-based techniques. The interaction of a customer in a retail environment is highly complex, but treated well using this method.

Sette and Van Langenhove present the use of efficiency based classification systems applied through genetic algorithms and fuzzy logic to model the fiber-to-yarn conversion process. This work contains a great deal of experimental data on the yarn spinning process and provides some insightful input into the problem of predicting yarn strength and quality with knowledge of processing parameters.

The intended audience of this book includes professionals, researchers and developers of software/hardware tools for the design of soft computing-based systems exploiting issues in textile sciences, and the entire computational intelligence community. It is expected that the reader is a graduate of textile engineering, computer engineering, or computer science study program with a modest mathematical background. Our book forms also a good basis for Ph.D. level seminars on soft computing in textile sciences.

Les M. Sztandera and Christopher Pastore
Philadelphia University
Philadelphia, Pennsylvania, U.S.A.

References

1. Zadeh L.A. (1981), Possibility theory and soft data analysis, in *Mathematical Frontiers of the Social and Policy Sciences*, Cobb L. and Thrall R.M. (Eds.), Westview Press, Boulder, CO, U.S.A., pp. 69–129.

"Everything should be made as simple as possible, but no simpler."

– Albert Einstein –

Contents

Chapter 1 Soft Computing for Softgoods Supply Chain Analysis and Decision Support

Shu-Cherng Fang, Henry L.W. Nuttle, Russell E. King, and James R. Wilson

Department of Industrial Engineering and

Graduate Program in Operations Research

North Carolina State University, Raleigh, NC 27695-7906, USA

Summary: Research on soft computing techniques for decision support for the design and management of the softgoods supply chain are presented. In particular, this work has been directed to creating and demonstrating a fuzzy-neural soft computing framework for supply chain modeling and optimization and creating and demonstrating soft computing based approaches to capacity allocation and delivery date assignment. The former has required the development of fuzzy system identification procedures, a method for constructing membership functions for fuzzy sets, and a flexible supply chain simulation capability. The paper gives an overview of this work and the prototype tools we have developed.

1 Introduction

The term "supply chain" has been used since the 1980s to describe the whole spectrum of operations in almost every manufacturing industry; from purchasing of raw material, through transformation production processes, to distribution of the finished inventory to customers. As the complexity increases a supply chain is well depicted as a network of suppliers, manufacturers and customers. In the softgoods industry the overall supply chain includes fiber, textile, cut and sew, retail, and consumer. Figure 1 shows an overview of the softgoods pipeline and Figure 2 illustrates a general structure of such a supply chain. A description of detailed simulation models for the individual entities in the apparel supply chain can be found in [1,2].

A softgoods supply chain involves the activity and interaction of many entities. Usually each of these entities knows how to make locally optimal decisions when the situation is clear. Unfortunately many decisions must be made in settings involving vagueness and uncertainty. Furthermore successful supply chain operation requires coordination of the decisions of the individual entities while the level of uncertainty is amplified as information is passed through the chain. Even in the emerging data rich environment with current information technology (EDI, Internet, data mining), lack of fundamental knowledge about supply chain operation in a vague and uncertain environment is still a key problem faced by the industry. Understanding capacity/cost tradeoffs and coordinated operation of a softgoods supply chain operating in a vague and uncertain environment is essential for success in the highly competitive global market. This paper demonstrates the use of fuzzy mathematics, neural networks, and other soft computing technologies in addressing critical softgoods supply chain integration and decision support problems.

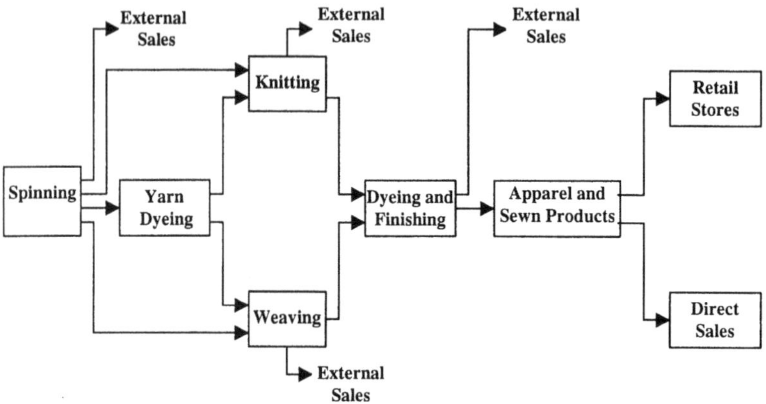

Figure 1. Overview of the softgoods pipeline

To date there has been no rigorous theoretical treatment of supply chain operation in vague and uncertain environments. Nor are there reliable, fully disclosed, science-based decision support tools. Existing approaches for coordinating the activities in a supply chain require the specification of precise quantities such as capacity levels and customers' desired delivery dates. However the true nature of the problem involves data and objectives which are often vague and imprecise.

For example, many customers of an apparel manufacturer will be able to tolerate delivery somewhat later than their nominal order due-date. Thus order due-dates are somewhat flexible (vague). The manufacturer has a "fuzzy capacity" in that there are options to schedule overtime, subcontract locally, or even go offshore. Management wants a “high” level of service but at the same time “low” inventories.

In order to make good decisions the apparel manufacturer needs to coordinate local activities with those of upstream suppliers and downstream customers- with uncertainty and imprecision present on all fronts. Other entities in the chain are faced with a similar problem. The coordination of numerous activities, particularly when different firms are involved, requires negotiation and compromise. This requires an approach which can be flexible enough to accommodate imprecise linguistic data as well as precise numerical data and which yields solutions that will provide compromise among different parties' objectives.

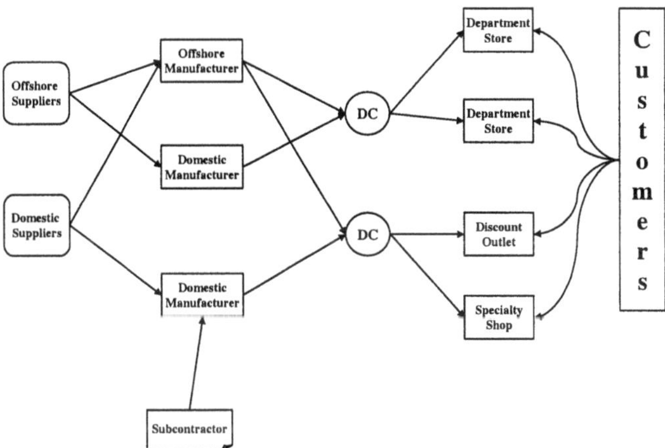

Figure 2. General supply chain architecture

To provide intelligent, responsive knowledge for decision support which can accommodate these characteristics, we have developed and prototyped a soft computing framework for supply chain modeling and optimization. In conjunction with this activity, we have created a flexible supply chain simulation capability, developed an efficient approach for constructing the membership functions needed to model imprecise quantities with fuzzy sets and developed and tested new procedures for knowledge extraction from operational data. We have also conceived and prototyped several versions of decision support tools for interactive due-date negotiation.

2 Supply Chain Modeling and Optimization Using Soft Computing Based Simulation

In order to provide a vehicle for softgoods supply chain modeling, analysis, and optimization incorporating the uncertainty and imprecision inherent in real systems, we have developed a soft computing guided simulation system.

While simulation can help the decision maker to understand better the supply chain, many different combinations and lines of action are possible to improve the whole system. It is typical that the simulation analysts and experts have to spend a considerable amount of time trying to change the original system searching for a good design and balancing several conflicting objectives simultaneously. This trial and error procedure can be avoided by coupling soft computing techniques, including fuzzy logic, evolutionary programs and neural networks, with the simulation of the supply chain.

A schematic of the soft computing guided simulation approach is given in Fig. 3.

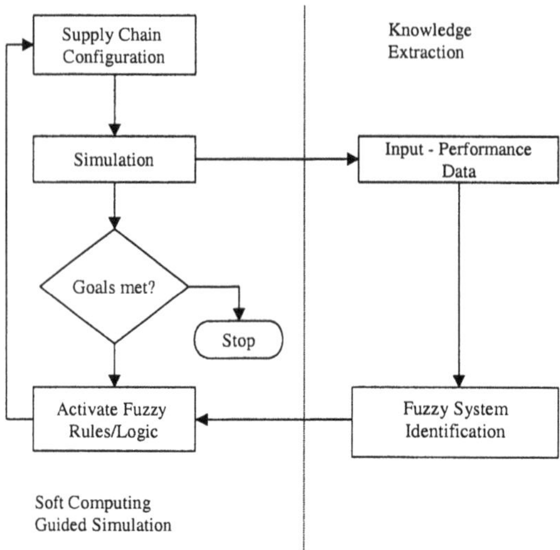

Figure 3. Soft computing guided simulation system

The system has two major components, the soft computing guided simulation procedure (on left) and a knowledge extraction procedure (on right). The simulation procedure is executed iteratively, beginning with a supply chain structure (manufacturers, suppliers, customers, etc.), a specific set of operational parameter settings (inventory levels, production capacities, lead-times, etc.) and specific management goals (such as "we want customer service to be HIGH and inventories to be LOW"). The operation of the system is simulated for a period of time and performance measures calculated. Observed performance is then compared with stated goals. If the supply chain objectives are not yet achieved, the system will check with its fuzzy knowledge base having its latest system performance measures on hand. After this dialog, fuzzy rules contained in the knowledge base will be activated to adjust parameters in the simulation model. This process is repeated until the system objectives are met to a high degree. The theoretical foundation of the proposed approach can be found in [3,4].

The knowledge extraction procedure is used to create the initial rule base and/or revise a current rule base based on observed simulation results. Three of the key components of the overall system are the simulator, the fuzzy system identification procedure, and a mechanism for constructing fuzzy set membership functions. These are described in more detail in the next three sub-sections.

To test the validity of our approach, a simulation model for a simple four-stage supply chain such as that illustrated in Figure 4 was created. Each stage of the chain has parallel processing units and limited inventory buffer capacity. The controllable parameters are the number of processing units and buffer capacity at each stage.

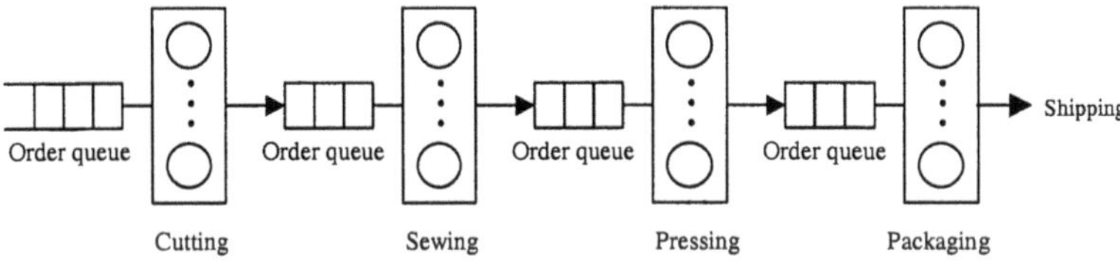

Figure 4. Simple four-stage supply chain

The graph in Figure 5 illustrates how the soft computing guided simulation system is able to quickly adjust supply chain parameters to obtain settings yielding a HIGH customer service level in very few iterations. More detailed results can be found in [4].

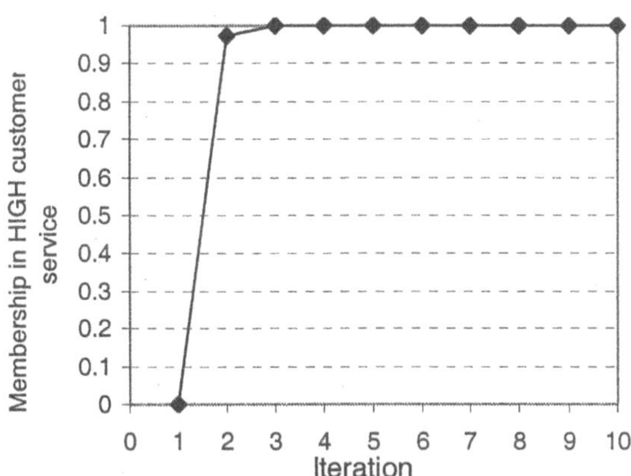

Figure 5. Control path for overall work-in-process

2.1 Supply Chain Simulator

In order to quickly create a flexible supply chain simulation capability, we have developed an interactive simulator written in Visual C++ with an interface as shown in Figure 6.

With this simulator the user configures an existing or contemplated supply chain, specifying the entities, their connectivity, the product line(s) and associated bills of materials. Entities are placed in the display using a simple drag-and-drop feature. The generic entities include customers, retailers, distribution centers, manufacturers, and suppliers. The bills of materials are entered into an ACCESS database.

The operational characteristics of each entity are specified by entering a number of parameter values in a dialogue box which is opened by clicking on the respective entity. The information entered includes demand characteristics, inventory control and reorder policies, production capacities, order lead-time and fill-rate characteristics, and various cost parameters. Figure 7 illustrates the parameterization of the cut and sew operation in the supply chain in Figure 6. Using five tabs, the user specifies the parameters which govern production process, inventory control of product and components, shipping of product and ordering or components. Performance goals for the chain are entered in a similar manner.

We note that this simulator is proving to be useful in its own right, independent of the soft computing guided system. In addition to textile industry applications, we are currently using it as the basis for a supply chain design tool for the furniture industry.

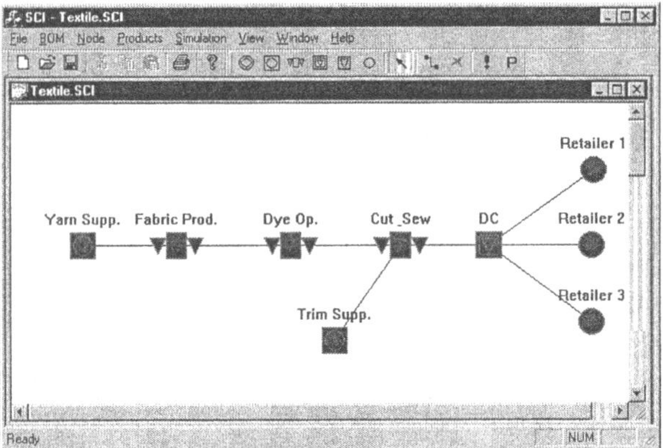

Figure 6. Supply chain simulator interface

2.2 Knowledge Extraction from Simulated Operational Data

To identify underlying system dynamics in order to generate (or modify) the fuzzy "if-then rules" used to guide the operational parameter adjustment, we have developed and tested new methods for extracting knowledge from input-performance data from an (in this case simulated) operational system.

"System identification" involves identifying that model within a class which may be regarded as equivalent to a target operational system with respect to input-performance data pairs. The identified model can then be used to explain and modify the behavior of the target system. In our case the target system is the (hopefully small) set of rules which will enable rapid operational parameter adjustment in the simulated supply chain to provide a high level of satisfaction of stated performance goals.

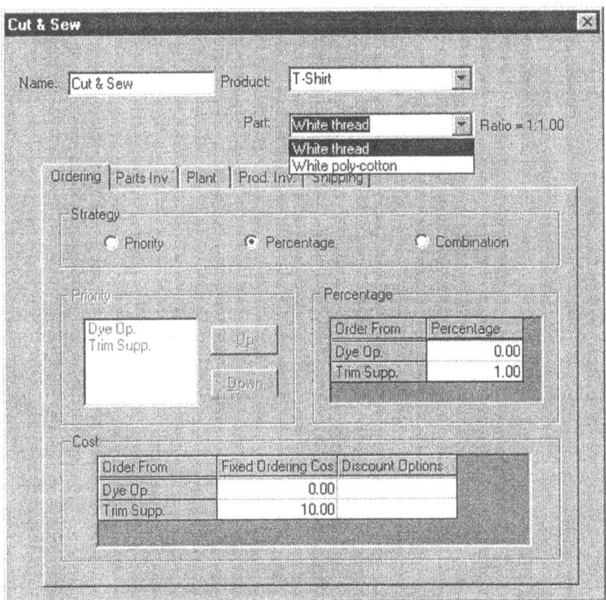

Figure 7. Dialogue box for operational parameter setting

Our first approach is based on Takagi and Sugeno's seminal work [5]. It consists of two phases. The first phase provides a baseline fuzzy model of the operational system. This is implemented by integrating the subtractive clustering method with the fuzzy c-means clustering algorithm. The second phase uses steepest descent and recursive least-squares estimation methods to fine tune the parameters of the baseline design to provide a better match with the target system. This approach has been able to successfully identify small sets of rules which provide a high level of performance on test problems from the literature.

Since Phase 2 turned out to be computationally slower than hoped, we examined an alternative clustering approach for Phase 1 with the objective of reducing the required Phase 2 effort. This proved quite successful. In fact with the same test problems the second phase was not required at all in order to achieve comparable performance. The theory and experimental results of the proposed two phase approach to fuzzy system identification can be found in [6,7]. With any given set of simulated operational data, the proposed two-phase method quickly extracts the knowledge embedded in the data set by identifying a family of control rules in the form of "*if* situation is this, *then* do that". These "if-then" rules are then activated by the fuzzy logic to adjust the parameter values for the desired performance.

In addition to using fuzzy clustering methods we have done some preliminary work in applying neural network technology to provide the fuzzy rule base. Application to the same test problems suggests that the neural network approach can provide rules with a higher level of performance based on small amounts of training data. However, according to our experiments with additional data, the clustering-based approach's performance rises to exceed that of the neural network approach.

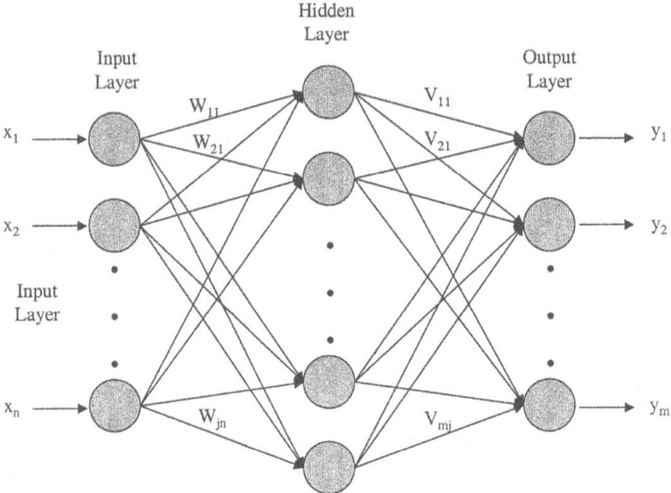

Figure 8. Neural network architecture

We have also used neural networks to develop a meta-model of the relationships between key input parameters and performance measures of a given operation. A three-layer network using the logistic function for activation with back propagation structure shown in Figure 8 is adopted for our studies. To speed up the learning process, an efficient neural network learning rule using the second order information with a fuzzy controller has been recently proposed in [8]. Application to the textile spinning operation can be found in [9] while application to apparel retail operations can be found in [10]. Additional information can also be found in [10]. The resulting neural networks are used as the engine in an interactive, graphical management information system which is included in a software package called the Sourcing Simulator which is available from $[TC]^2$ (Textile Clothing and Technology Corporation). Figure 9 shows the decision surface model of the relationship between customer "service level", replenishment "lead-time", and beginning of selling season " inventory" from a Sourcing Simulator scenario.

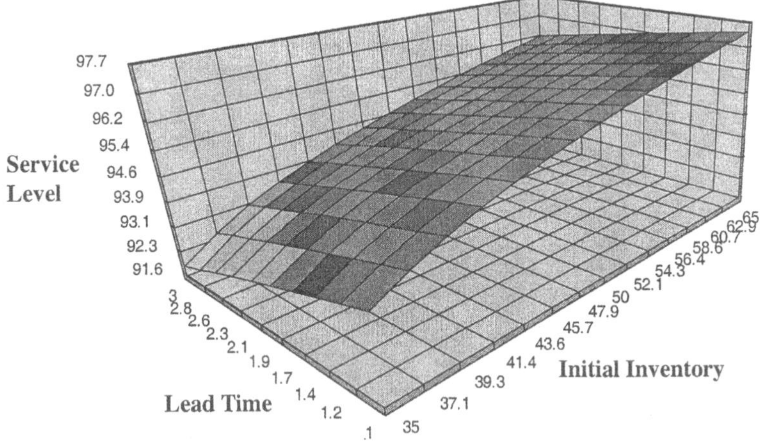

Figure 9. Example neural network decision surface plot from Sourcing Simulator

2.3 Membership Function Construction

In our system depicted in Figure 3, imprecise concepts such as HIGH customer service and LOW work-in-process inventory level are modeled as fuzzy sets. Figure 10 illustrates a possible fuzzy set representation of MEDIUM machine utilization. In this case, utilization levels around 50% are regarded as definitely "medium" and thus have membership values at or close to 1. On the other hand utilization levels below 10% and above 75% are definitely not "medium" and thus have membership values of 0. Points in between have memberships which rise toward 1 the closer they are to 50%.

In current practice, modelers choose the shape of the membership function from a pool of commonly used parameterized families including triangular, trapezoidal, Gaussian, sigmoid, and S-shaped. After a shape is selected, the parameters are manipulated to tune the shape. In contrast, we have developed an approach which employees Bezièr curves which, with the aid of control points (the black dots in Figure 10), can be used to produce the membership of almost any imprecise concept. The underlying mathematical problem involved is to solve a mixed integer nonlinear program. This new methodology with computational experiments has been reported in [12]. It has the ability to fit any given data set with a minimum level of discrepancy. Moreover, in the absence of data, the methodology can be intuitively manipulated by the users to construct membership functions with the desired shape by changing the position of control points.

Intuitively, each control point acts like a magnet which attracts the membership function to bend toward the control point.

This new flexible and interactive way of building and tuning membership functions can be leveraged by using a graphical user interface (GUI). We have developed a GUI that helps the modeler add, move, delete control points to obtain the desired membership function.

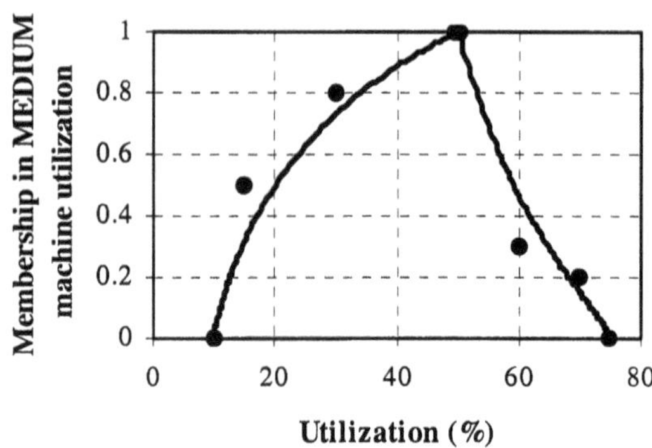

Figure 10. Membership function for MEDIUM machine utilization

3 Due-Date Negotiation

How to negotiate order due-dates that are acceptable to both a manufacturer and its customers is an important issue in the make-to-order manufacturing systems, since sales normally depend on both the cost and delivery date. Traditionally, a customer negotiates a required due-date with a salesperson who relies on the *sales management* module of a Manufacturing Resource Planning (MRP-II) system. However, since the sales management module is not normally linked with the *production planning* module of the MRP-II system, the salesperson is not able to get detailed information relative to the availability of various manufacturing resources. Therefore, in practice, a customer tends to ask for the earliest possible due-date, and, to get the order, a salesperson tends to promise the customer a due-date without adequate consideration of the availability of production capacity. This often results in tardy deliveries, unhappy customers, and low utilization of manufacturing facilities. Figure 11 shows our new scheme for due-date bargaining proposed in [13], which provides integration between the sales management and production planning functions.

Since Supply Chain Management (SCM) first attracted the attention of researchers and managers, a number of commercial software packages have been developed and implemented in actual manufacturing enterprises. Although some packages include functions, such as ATP (Available-To-Promise) and CTP (Capacity-To-Promise) to support the manufacturer's order acceptance/rejection decision, how to support the negotiation between manufacturers and their customers has not been properly addressed from either an academic or practical perspective.

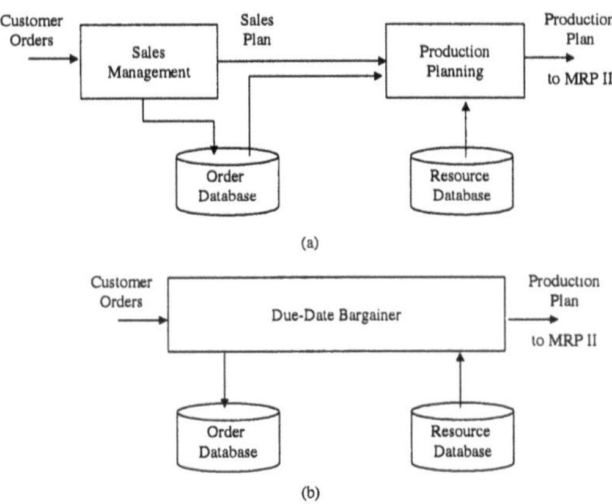

Figure 11. Comparison of management approaches

To support such negotiation, we have developed and prototyped two approaches. The first is a due-date bargaining method which uses fuzzy modeling to capture the imprecision inherent in "shop capacity" and customer specified due-dates in terms of tolerance level. A mixed integer fuzzy linear programming model is used to allocate fuzzy shop capacity to meet customer specified due-dates. Because of the mixed integer variables, the underlying solution procedure for this fuzzy optimization model uses genetic algorithms along with fuzzy logic for an optimal capacity allocation. When the allocation misses a customer's desired due-date, the customer can bargain for an earlier due to be met at a cost. Capacity reallocation as the result of the bargaining process may continue as longer as desired. The details can be found in [13,14,15].

For testing and demonstration we have implemented the method in a prototype computer software package which is oriented to apparel manufacturing enterprises. We call it the "Multi-Customer Due-Date Bargainer." The package consists three modules, Input Data Management, Due-Date Assignment, and MPS Management. Input Data Management provides an interface (Figure 12) for data entry/editing. Data on customers, products, orders, bill of materials manufacturing resources (e.g., cutting, sewing, pressing, packaging), and shop calendar can be viewed and edited on one of five tabs on an input form.

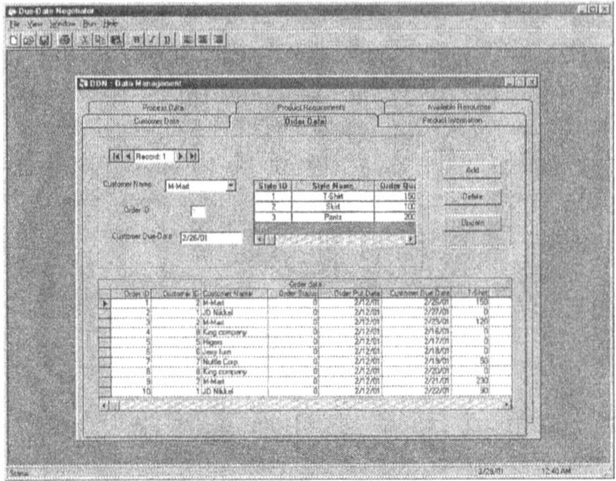

Figure 12. Input Data Management interface

The Due-Date Assignment module contains the solver. A C++ "DLL" implements the procedure introduced above. MPS Management is the output module. It provides GANTT charts showing the current loading of individual orders to the various resources and the associated resource load profiles (Figure 13).

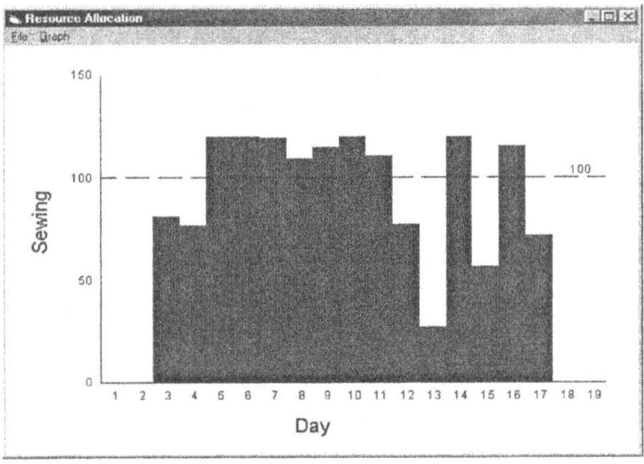

Figure 13 Example resource profile from Due-Date Bargainer

The second approach to support due-date negotiation provides the manufacturer with greater flexibility in exploring alternatives. A real-time due-date assignment approach is combined with MRP-II based on the concept of integrating the due-date assignment process with the production planning process. Potential new customer orders are inserted into a rough-cut capacity plan which details the implied time–phased work load on each key resource and the associated estimated order completion dates. First, leaving the plan for currently active orders undisturbed, earliest possible completion times for new orders that do not overload production resources are determined. If the resulting estimated completion times satisfy the customers' requested delivery dates, the order promise dates can be quoted as requested. However, in many instances some of these estimated completion dates may not meet customers' requirements. In this case the prototype software allows the manufacturer to determine the impact on the plan selectively scheduling overtime on one or more resources and/or of forcing the loading of one or more orders to meet specific delivery dates. Exploring a number of options permits the manufacturer to make informed delivery date quotations.

While exploring such alternatives, the loading for selected customer orders can be left undisturbed. This version does not use fuzzy logic; however, each execution of the loading logic uses a genetic algorithm in a effort to find that loading sequence for schedulable orders with the best requested delivery date achievement.

Figure 14 illustrates the screen that is presented to the user upon the completion of a loading run. The due-dates in the second column are those requested by the customer while the dates in the last column are those obtained with the last loading run. The "Schedule Option" section indicates that another loading run will be made in an effort to obtain the specified target due dates for orders 3 and 10, and in doing so the dates for four orders will be fixed at the current values while those for the remaining three orders may be adjusted as necessary.

DUE DATE ASSIGNMENT/SCHEDULER

Order ID	Customer Due-Date	Schedule Option	Target Due-Date	Promised Due-Date
1	18-Jul-00	Fixed		18-Jul-00
2	15-Jul-00	Fixed		15-Jul-00
3	15-Jul-00	Negotiating	15-Jul-00	16-Jul-00
8	11-Jul-00	Free		12-Jul-00
9	12-Jul-00	Fixed		13-Jul-00
10	13-Jul-00	Negotiating	13-Jul-00	17-Jul-00
11	24-Jul-00	Free		24-Jul-00
12	14-Aug-00	Fixed		14-Aug-00
13	11-Aug-00	Free		11-Aug-00

Figure 14. Due-date achievement report

4 Conclusion

We have developed and prototyped a soft computing framework for the softgoods supply chain modeling and optimization. In conjunction with this activity, we have created a flexible supply chain simulation capability, developed an efficient approach for constructing the membership functions needed to model imprecise quantities with fuzzy sets and developed and tested new procedures for knowledge extraction from operational data. We have conceived and prototyped several versions of decision support tools for interactive due-date negotiation. Our earlier neural network based decision surface modeling tool is now included in Version 2.0 of the DAMA project's Sourcing Simulator distributed by $[TC]^2$.

5 Acknowledgement

This work has been supported by the National Textile Center. Other contributors include our current and former graduate students: Shyh-Huei Chen, Hao Cheng, Ta-Wei Hung, Saowanee Lertworasirikul, Yi Liao, Andres Medaglia, and Peitang Wu at NC State University.

6 References

1. Nuttle, H.L.W., R.E. King, N.A. Hunter, J. R. Wilson, and S.-C. Fang, "Simulation Modeling of the Textile Supply Chain, Part I – The Textile Plant Model," to appear in *The Journal of the Textile Institute*, 2001.

2. Nuttle, H.L.W., R.E. King, S.-C. Fang, J. R. Wilson, and N.A. Hunter, "Simulation Modeling of the Textile Supply Chain, Part II - Results and Research Directions," to appear in *The Journal of the Textile Institute*, 2001.

3. Medaglia, A.L., S-C. Fang and H.L.W. Nuttle, "Fuzzy Controlled Simulation Optimization," to appear in *Fuzzy Sets and Systems,* 2001.

4. Medaglia, A.L., "Simulation Optimization Using Soft Computing," PhD dissertation, North Carolina State University, Graduate Program in Operations Research, Raleigh, NC, 2000.

5. Takagi, T. and M. Sugeno, "Fuzzy Identification of Systems and Its Applications to Modeling and Control," IEEE Transactions on Systems, Man, and Cybernetics, Vol. SMC-15, 116-132, 1985.

6. Hung, T-W, S-C. Fang, and H.L.W. Nuttle, "A Two-Phased Approach to Fuzzy System Identification," under review by *Fuzzy Sets and Systems*, 1999.

7. Hung, "A New Approach to Fuzzy System Identification," PhD dissertation, North Carolina State University, Graduate Program in Operations Research, Raleigh, NC, 1999.

8. Wu, P., S-C. Fang, and H.L.W. Nuttle "Efficient Neural Network Learning Using Second Order Information with Fuzzy Control," NEUCOM 1230 to appear in *Neurocomputing,* 2001.

9. Wu, P., S-C. Fang, H. L. W. Nuttle, R. E. King, and James R. Wilson, "Guided Neural Network Learning Using a Fuzzy Controller with Applications to Textile Spinning," *International Transactions in Operational Research,* 2, No. 3, 259-272, 1995

10. Wu, P., S.-C. Fang, H.L.W. Nuttle, and R.E. King, "Decision Surface Modeling of Apparel Retail Operations Using Neural Network Technology," *International Journal of Operations and Quantitative Management,* 1, No. 1, 33-48, 1995.

11. Wu, P., "Neural Networks and Fuzzy Control with Applications to Textile Manufacturing and Management", Ph.D. Dissertation, Graduate Program in Operations Research, North Carolina State University, Raleigh, NC, 1997.

12. Medaglia, A.L., S-C. Fang, H.L.W. Nuttle, and J.R. Wilson, “An Efficient, Flexible Mechanism for Constructing Membership Functions”, to appear in *European Journal of Operational Research,* 2001.

13. Wang, D-W., S-C. Fang, and T.J. Hodgson, "A Fuzzy Due-Date Bargainer for Make-to-Order Manufacturing Systems," *IEEE Transactions on Systems, Man, and Cybernetics,* 28, No. 3, 492-497, 1998.

14. Wang, D-W., S-C. Fang, and H.L.W. Nuttle, "Soft Computing for Multi-Customer Due-Date Bargaining", *IEEE Transactions on Systems, Man, and Cybernetics,* Vol. 29, No.4, 1999.

15. Wang, D-W., S-C. Fang, and H.L.W. Nuttle, Fuzzy Rule Quantification and Its Application in Manufacturing Systems", *Journal of Chinese Institute of Industrial Engineering* (Special Issue on Softcomputing in Industrial Engineering), Vol. 17, No. 5, 505-516.

Chapter 2 Application of Fuzzy Set Theory in Mechanics of Composite Materials

A. Muc, P. Kedziora

Institute of Mechanics & Machine Design, Krakow University of Technology,

ul. Warszawska 24, 31-155 Kraków, Poland

Keywords:Fuzzy Sets, Mechanical Properties, Buckling, Delamination, Fatigue, First-Ply-Failure, Optimization

1 Introduction

1.1 Composites - Limits of the Deterministic Description

The field of composites mechanics, engineering and technology is relatively young, and the test methods and measurements techniques are not yet fully developed. The modeling of their mechanical properties is even further behind the experimental investigations. The study and application of composite materials is a truly interdisciplinary endeavor that has been enriched by contributions from chemistry, physics, material science and manufacturing engineering. Since numerous possibilities exist in combining constituents to form a composite there are many factors

that can affect the global homogenized mechanical properties of composite materials, their behavior under different boundary and loading conditions, and their final failure. In fiber composites, both the fibers and the matrix retain their original physical and chemical identities, yet together they produce a combination of mechanical properties that cannot be achieved with either of the constituents acting alone, due to the presence of an interface between these two constituents. Thus, proper characterization of composites, whether it is for chemical, physical or mechanical properties, is extremely difficult because most interfaces are buried inside the material.

The description of damage mechanisms with respect to short-term strength and long-term static fatigue conditions requires much experimental data dealing with the behavior of not only constituents in composites but also of the interfaces as well as defects and/or internal strains and stresses arising during manufacturing regime. The problems of failure have been extensively studied both theoretically and experimentally, and considerable data has been carefully generated. The classic works [1] – [6] have resulted in a fundamental understanding of the chemical and kinetic mechanisms involved, including the effects of the stress corrosion in humid and alkali environments. Kies [7], Shmitz and Metcalfe [8, 9] carried out extensive studies of the flaw types, their sources during processing and handling and especially their statistical distributions. Charles and Hillig [6] further advanced the understanding of the stress dependence in surface reactions and how flaws geometrically evolve from cracks.

Based on the literature we believe it is possible to build a mechanistic model of the failure under steady and cyclic stresses that is much more advanced and comprehensive than has been attempted to date. Such a model should incorporate effects of temperature, alkali (and acid) species and concentrations, probability distributions shapes, and scaling laws for lifetime (scale parameter or mean) versus stress level. This goal of building a comprehensive model will be accomplished by incorporating: (i) fiber mechanical properties and flaw statistics at the length scale of local fiber load transfer, including any stochastic stress-time dependency, (ii) matrix plasticity and creep in shear around fiber breaks, (iii) interface dis-bonding and viscous frictional sliding in terms of local shear stress, (iv) local fiber and matrix packing geometry (2-D planar, 3-D hexagonal and random) and associated residual stresses from processing, and (v) multiple matrix cracking as it affects penetration of the environment to the fiber surface.

The experience with modeling of composite materials properties and damage shows clearly that the knowledge of the relation between the microstructure and the macroscopic mechanical properties of a material is a prerequisite for modern material design. Many complex microstructural phenomena constitute the basis of nonlinear mechanical behavior in general and of non-local material behavior in particular. The introduction of non-local models in nonlinear mechanics was initially introduced as a remedy for the ill-posed boundary value problem that is a consequence of softening behavior. However, theoretical analyses and experimental investigation have shown that non-locality has a physical nature that is closely related to the microstructure of the material through the incorporation of an intrinsic length scale. In this area, the present research focuses on the relationship between the length scale and certain microstructural aspects in composites. The experimental validation of a constitutive model is an important issue in computational mechanics.

It is important to emphasize that the combined use of sophisticated measurement techniques, computational simulations and a numerical identification tool for the model parameters may result in a powerful method for material characterization of mechanical properties and damage analysis. However, in such an analysis the knowledge of a great number of various factors is required and the validity of a pure deterministic modeling is always not complete due to the scatter of experimental data. Thus, many engineering problems connected with the use of composite materials are (or may be) too complex and ill defined to be modeled by conventional deterministic procedures. On the other hand, the use of statistical analysis is also limited to: (1) the number of experimental data (the extension of random fields of variables), (2) the definition of covariance matrices since the majority of random variables are correlated in an arbitrary manner, and (3) computational efforts in the analysis of multidimensional statistical (stochastic) problems.

The monograph written by Elishakoff [10] presents the variety of problems that one can encounter in the statistical approach but in the case of composite material properties or damage the scale of problems increases significantly due to the anisotropy and complexity of the materials. Currently woven fabric composites are widely used in technical practice. The variables under control include the fiber and resin type, the yarn type and the weave type. It is important to understand how the properties of the constitutive phases and the geometric characteristics of the fabric combine to determine the properties of the final composite materials. This is a complex task since it requires a computational 3-D analysis. However, such an investigation is conducted on the idealized models only that are based on the analysis of a regular, periodic array of material unit cells. Therefore, this is an additional (to the mentioned above problems) origin of uncertainties in the description of composite materials.

1.2 Descriptions of the Uncertainties

Referring to composite materials, the origin and source of imprecision (or uncertainty) lies mainly in the lack of information dealing with their microstructure, mechanical properties, behavior and the number of factors responsible for gradual degradation of their properties and final failure. Commonly, the theoretical (deterministic) analysis of composites is based on homogenization theories that may include an increasing number of different parameters. However, it is unknown in advance what number of parameters is sufficient to describe satisfactorily the problem considered. On the other hand, the material parameters are evaluated in the experimental way being the source of randomness in the traditional (deterministic) analysis or impreciseness or vagueness in the fuzzy set approach. The imprecise, vague, qualitative, linguistic or incomplete information may be present in geometry, material properties, degradation of properties, applied loads or boundary conditions.

In the analysis of engineering problems three different approaches are used depending on the nature and extent of uncertainty by the introduction of the uncertainty triangle (Rao and Sawyer [11]) – see Fig. 1. If we know the probability distributions of the system variables and the covariance matrices the performance of the system may be considered with the use of probability theory. If only the fragmentary information on the uncertainty quantity is available an upper bound on the maximum response may be only evaluated using the anti-optimization approach.

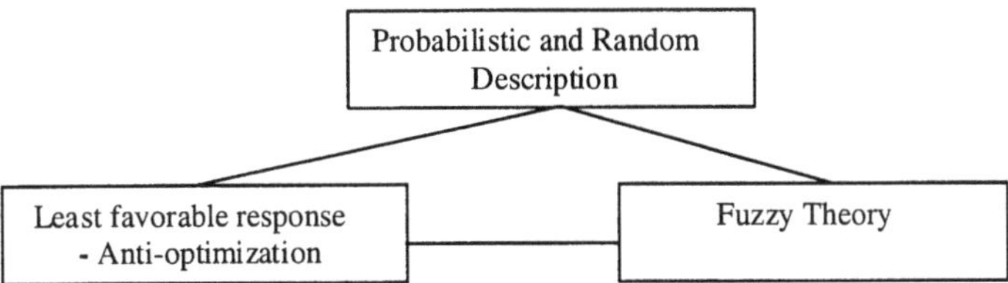

Figure 1. Approaches to modeling uncertainty

If the system parameters are described in imprecise or linguistic terms the fuzzy theory can be implemented to predict the structural response in the sense of evaluation of its upper and lower bounds, respectively.

1.3 Objectives of the Present Work

Briefly speaking, the present work is devoted to the discussion of various problems associated with the description of different phenomena encountered in the mechanics of composite materials via the fuzzy set theory.

In the first part a theoretical background of the fuzzy set theory is explained. First the differences between fuzzy numbers and fuzzy qualifiers are pointed out. Special attention is focused on the construction of the membership functions and then on modeling using these functions in a computational strategy for the evaluation of lower and upper bounds of the structural response. The problems of using various experimental data in the approximations of membership functions and the building of the Fuzzy Knowledge Base are also presented.

The second part of the chapter deals with the demonstration of the application of the fuzzy set approach in various mechanical problems. In detail, the analysis exhibits: (i) definitions of material properties for unidirectional and textile composites (the application of micromechanical models), (ii) damage analysis of the limit load carrying capacity of composite structures including buckling response, the first-ply-failure (FPF) and fatigue problems and (iii) stacking sequence (topology) optimization of composite structures in a fuzzy environment.

The discussion and illustration of different problems is also supplemented by the review of the literature available in this area although we do not pretend to present all references dealing with the use of the fuzzy set approach in engineering, mechanics and mechanics of composites.

2 Foundations of Fuzzy Set Theory

This section deals with the foundations of the fuzzy set theory. First we present the notation of a fuzzy set and discuss forms of their possible representations. Then we present some aspects of the fuzzy set theory applications in view of a numerical, theoretical and experimental analysis.

2.1 Fuzzy Logic – a Brief Literature Review

Fuzziness is property of language. Its main source is the imprecision involved in defining and using symbols, processes or physical phenomena. As information dealing with physical phenomena is vague, imprecise, qualitative, linguistic or incomplete, the problem cannot be formulated in a classical, standard (i.e. deterministic or statistical) way.

Although fuzzy logic has been described and examined for nearly thirty years, it has only recently appeared in the popular and technical press. The commercial interest in fuzzy logic was quite low for a long time after it was first introduced by Lofti Zadeh [12] in 1965. However applying fuzzy logic to engineering problems represents only a fraction of its real potential. As the complexity of the target system increases a corresponding rapid decline in information afforded by traditional mathematical models is observed. Variables used by the models can only represent the state of a phenomenon as either existing or not existing. The general idea of the above describe the Zadeh "*principle of incompatibility*":

As the complexity of a system increases, our ability to make precise and yet significant statements about its behavior diminishes until the threshold is reached beyond which precision and significance (or relevance) become almost mutually exclusive characteristics.

The idea of fuzzy sets arises from the definition: fuzzy sets are functions that map a value, which might be a member of a set, to a number between zero and one indicating its actual degree of membership.

Recent applications of the fuzzy set theory in various scientific areas, such as artificial intelligence, image processing, biological and medical sciences, economics, geography, sociology, psychology have indicated that the theory is indeed a useful tool for the quantification of impreciseness and vagueness in many problems. Most engineering applications of that theory have been related to controls, decision-making and optimization.

It is beyond the scope of this chapter to provide a detailed review of the relevant literature, and the reader is referred to a few representative monographs: McNeill, Thro [13], Cox [14], Tsoukalas, Uhrig [15], Dubois, Prade [16], Zadeh [17], Kosko [18], where various problems connected with fundamentals and concepts used in the fuzzy set theory are presented. It is worth to mention also that a variety of numerical programs describing concepts of the fuzzy set theory are now available, e.g.: *MATHEMATICA, MATLAB* or the libraries of the procedures [19,20] that can be easily adopted in the *VISUAL BASIC* or *VISUAL C++* packages.

Generally, the existing works characterizing the applicability of the fuzzy set theory in various areas of mechanics may be classified into a number of categories. One group of the analyses is devoted to the introduction of fuzzy approaches to the numerical methods understood in the classical sense as finite elements (Lallemand *et al.* [21]), boundary elements (Skrzypczyk, Burczynski [22]) or proposed the use of new fuzzy elements (Rao, Sawyer [11]). Fuzzy concepts are used in the description of failure problems of different composite materials (Abdel-Tawab, Noor [23], Muc, Kędziora [24]). Another class of analyses in a fuzzy set environment is concentrated on stability problems of different structures - Refs [24,25]. The application of fuzzy sets to several civil engineering problems was reviewed by Brown, Yao [26], Valliappan and Pham [27]. The fuzzy analysis of critical vibration frequencies was conducted by Loskiewicz-Buczak and Uhrig [28], Kohonen [29]. The formulation for structural fuzzy optimum design as well as solutions of particular problems are presented in Refs [30] – [35]. It is worth emphasizing that a great number of works in this area are connected with stacking sequence optimization for various composite structures including composite pressure vessels.

2.2 Fundamental Definitions

If we consider a certain space, for instance the set IN of all integers, we generally describe data by defining subsets of the given space. In the space IN the feature "*less than 10*" is characterized by the set:

$$A = \{\ 1, 2, 3, 4, 5, 6, 7, 8, 9\ \} \subset \mathrm{IN} \tag{1}$$

Another representation of the data "*integer less than 10*" is the definition of the characteristic function Π_A in the following way:

$$\Pi_A : IN \rightarrow \{0,1\}$$

$$\Pi_A(\eta) = \begin{cases} 1, & \text{if less than 10} \\ 0, & \text{otherwise} \end{cases} \tag{2}$$

This yields the value 1 for each element of the space IN that belongs to the set A and the value 0 for each element that does not. The above representation is commonly called as a crisp set. However, this concept cannot be used directly as we intend to characterize the typical property for composite materials as e.g.: the failure of CFRP under tension occurs as the tensile strain ε_x is equal to the ultimate value 0.015 . The characteristic function of this set is depicted in Fig. 2a . A problem arises if the linguistic term "*the failure under tension*" has to be described. It is well known that from the micromechanical point of view the failure starts from micro-cracks in the matrix for the strain values much lower than the value 0.015. In addition, the value 0.015 is usually an average value characterizing rather a scatter of random values of macro-cracks appearing at the strain level 0.015. Therefore, for some specimens one can observe the final (macro-scale) failure as ε_x is equal to 0.0159 or to 0.0141. A possible solution to this problem is to generalize the definition of the characteristic function in a way that it yields values from the interval [0, 1] and not just the two values of the set $\{0, 1\}$. This leads to the notion of a fuzzy set.

A fuzzy set μ of X is a function that maps the space X into the unit interval, i. e.:

$$\mu : X \rightarrow [0, 1] \qquad (3)$$

The value $\mu(x)$ denotes the membership function of x to the fuzzy set μ. Fig. 2b shows (subjectively defined) a membership function of the fuzzy set μ describing the linguistic meaning of the term "*the failure under tension*". The use of fuzzy sets to formally represent vague data is often done in an intuitive way because in many applications there is no model that provides a clear interpretation of the membership degrees, although we want or we try to base on various experimental data.

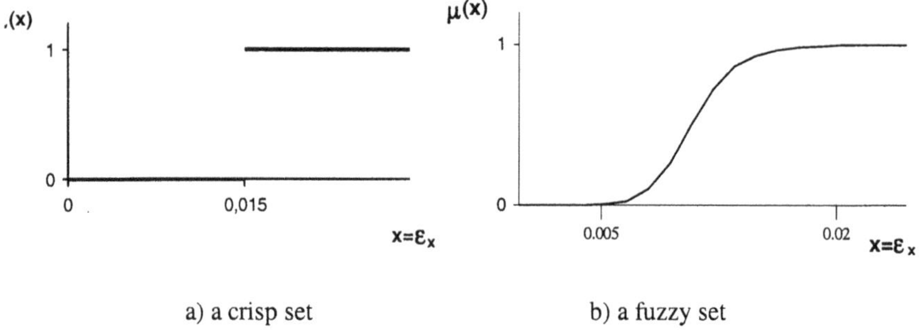

a) a crisp set b) a fuzzy set

Figure 2. The representation of the term "*the failure occurs* ..."

2.3 Representations of Fuzzy Sets – Membership Functions

The application of fuzzy methodologies requires knowledge of the membership functions of fuzzy quantities. In general, fuzzy numbers are sets that represent numeric quantity. It can be done in variety of ways. Of course, there are different possibilities to determine and represent membership functions characterizing a fuzzy set. If the subspace X contains only a finite member of elements, a fuzzy set μ of X will be defined by specifying for each element $x \in X$ its membership degree $\mu(x)$. If the number of elements is very large or a continuum is chosen for X then $\mu(x)$ can be better defined by a function that can use parameters that are adapted to the actual modeling problem. For instance, as we want to represent the term "*Young's modulus is equal to 200 [GPa]*" in the sense of a fuzzy set having a finite amount of the experimental data we can select one of different representations given in Fig. 3.

The membership function is interpreted as the measure of the compatibility between a value from the domain and the idea underlying the fuzzy set A, i.e.:

$$\mu_A(x) \leftarrow f(x \in A) \tag{4}$$

A fuzzy set A consists of three components: (1) a horizontal domain axis of monotonically increasing numbers that constitute the populations of the fuzzy set; (2) a vertical membership axis between zero and one, indicating the degree of the membership in the fuzzy set; (3) the surface of the fuzzy set itself that connects an element in the domain with the degree of membership in the set.

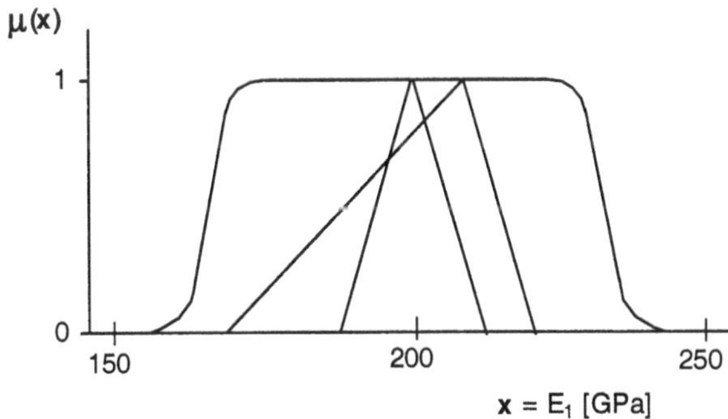

Figure 3. Various fuzzy representations of "*Young's modulus equals 200 [GPa]*"

This degree of membership is also known as the membership or truth function since it establishes a one-to-one correspondence between an element in the domain and the truth value indicating its degree of the membership in the set A. In this way we have combined imprecision with the ability to calculate a degree of membership at each point of the domain. This provides the basis for the calculus of fuzzy logic.

The surface of a fuzzy set , the part of the set that actually defines the membership function, can assume a wide range of shapes . The method of recognizing fuzzy attributes and drafting the fuzzy set is an important technique. However, when we turn to building fuzzy models, fuzzy systems are tolerant of approximations not only in their problem spaces but also in the representation of fuzzy sets. This means that they perform well even when the fuzzy set does not map exactly with the model concept.

There are, generally, two kinds of fuzzy sets:

- numbers that represent an approximate numeric quantity,
- qualifiers that characterize open-ended concepts; these sets provide the framework for describing unbounded concepts (or concepts that are theoretically unbounded).

We now examine how fuzzy sets are constructed and how they encode a particular kind of information.

Fuzzy numbers is an important class of fuzzy sets that are spread around a central value. The generalized approximation curve indicated by around or close produces a membership function for the fuzzy assertion representing a fuzzy space of all numbers that are around or close to Y:

$$\text{X is around Y} \tag{5}$$

In the composite materials field it means that any mechanical or geometrical property describing the material is not a crisp set.

In general, there are five important classes of fuzzy contours describing fuzzy numbers: (i) triangular curves, (ii) trapezoidal (shouldered) curves, (iii) the π (pi) curves, (iv) the β (beta) curves and (v) the Gaussian curves. They are defined in the following way:

triangular curves:
$$Tr(x;\gamma,\beta)=\begin{cases}\dfrac{x-\gamma+\beta/2}{\beta/2}, x\in[\gamma-\beta/2,\gamma]\\ \dfrac{\beta/2+\gamma-x}{\beta/2}, x\in[\gamma,\gamma+\beta/2]\end{cases} \tag{6}$$

— trapezoidal curves:

$$\text{Trap}(x;\beta,\gamma,\delta\} = \begin{cases}\dfrac{x-\gamma+\beta/2+\delta/2}{\beta/2} & , x\in[\gamma-\beta/2-\delta/2,\gamma-\delta/2]\\ 1, & x\in[\gamma-\delta/2,\gamma+\delta/2]\\ \dfrac{\gamma+\beta/2+\delta/2-x}{\beta/2}, & x\in[\gamma+\delta/2,\gamma+\beta/2+\delta/2]\end{cases} \tag{7}$$

— the π curves;

$$\Pi(x;\beta,\gamma)=\begin{cases}S(x;\gamma-\beta,\gamma-\beta/2,\gamma), & x\leq\gamma\\ 1-S(x;\gamma,\gamma+\beta/2,\gamma+\beta), & x>\gamma\end{cases} \tag{8}$$

the β curves:
$$B(x;\gamma,\beta)=\frac{1}{1+\left(\dfrac{x-\gamma}{\beta}\right)^2} \tag{9}$$

the Gaussian curves:
$$G(x;k,\gamma)=exp\left[-k(x-\gamma)^2\right] \tag{10}$$

where the S (sigmoid) curves are defined in the following way:

$$S(x;\alpha,\beta,\gamma)=\begin{cases} 0, & x\le\alpha \\ 2(x-\alpha)^2/(\beta-\alpha)^2 & \alpha\le x\le\gamma \\ 1-2(x-\beta)^2/(\beta-\alpha)^2 & \gamma\le x\le\beta \\ 1 & x\ge\beta \end{cases} \tag{11}$$

The curves are centered on a single value γ from the domain, whereas β (and δ for trapezoidal curves) are parameters that indicate the width of the curve's base.

Fuzzy qualifiers are used to approximate an unknown or poorly understood concept that is not a fuzzy number, for instance the concept of failure. In such a case there are two states – see also Fig.2. The first one corresponds to the lack of failure and has zero (or one, depending on the definition) membership and moves to the right with the values that have increasing (or decreasing respectively) set membership up to the value equal to one (zero, respectively). To describe fuzzy qualifiers two types of curves are commonly used: (i) linear curves or (ii) the sigmoid S curves, given by eq. (11) – see Fig. 4.

For composite materials, microscopic defects are coalesced and grow as micro-cracks in structural materials through the various loading histories. These defects and cracks cause a decrease in stiffness and reduction in strength. The detailed process of deterioration of the mechanical properties of materials is not so simple or clear. In macroscopic sense, it can be assumed that micro-cracks are continuously distributed in a material. The measure of distributions of micro-cracks may be assumed as an internal state variable that indicates the change of load carrying capacity of a material in continuum damage mechanics. The fact that the load carrying capacity or strength of materials changes means simply that the failure of a material has a fuzzy nature in the sense of fuzzy qualifiers. Therefore, the membership for failure is directly related to the damage variable both for static or fatigue loading conditions. Since in the continuum damage mechanics the stiffness or strength degradation are usually characterized by the Weibull distributions, the exponential–style membership function may be introduced herein in the following form:

$$\mu(x)=\begin{cases} exp[-(x/\beta)^{\alpha}], & x<X_t \\ \text{linear function}, & x\ge X_t \end{cases} \tag{12}$$

The determination of the membership functions is difficult as Norwich [36], Dombi [37] point out. Thus, the first attempt or trial in building the membership functions can be based on the statistical data – see e.g. the description of failure. However, it is to be noted that not all fuzzy quantities have statistical data for defining their membership functions. If statistical data exists the membership function can be determined as follows:

$$\mu(x) = \lambda\rho(x), \qquad \lambda = 1/\max[\rho(x)] \tag{13}$$

where $\rho(x)$ is a probability density function or its estimate derived from the histogram of the feature x used for defining fuzzy set.

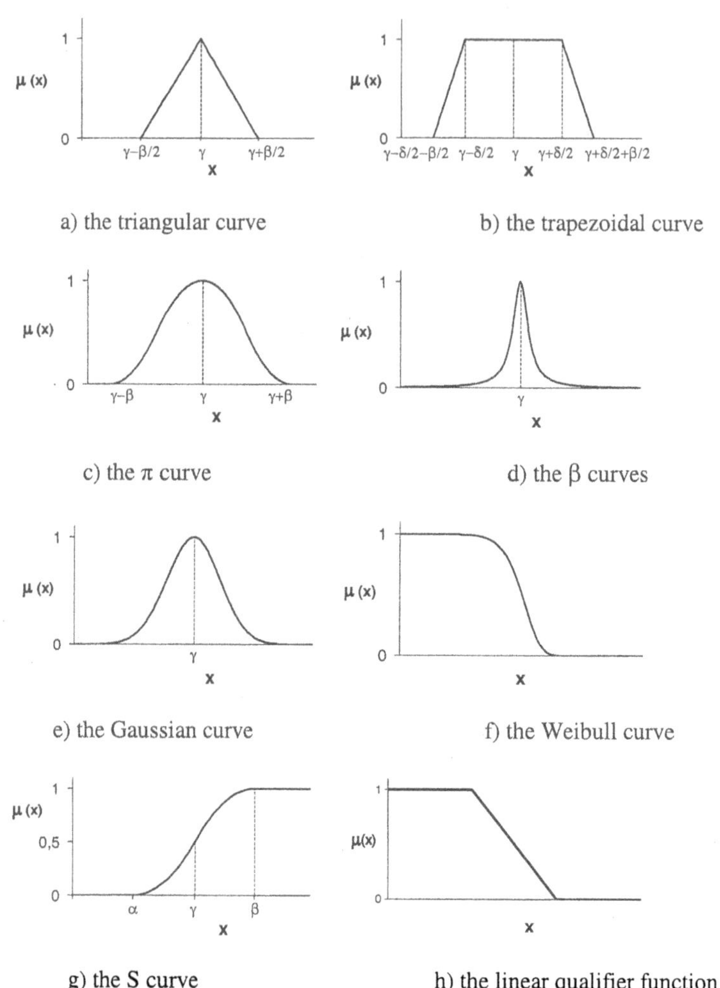

Figure 4. Shapes of typical membership functions – possible functional representations of fuzzy numbers (the curves a) – e)) and of fuzzy qualifiers (the curves f) – h))

It should be emphasized also that in the S curve the continuous cumulative distribution function $F(y)$ for the underlying concept variable takes the form of an S curve:

$$F(y) = Prob(x \leq y) \tag{14}$$

where y is a value from the underlying the domain and x is a randomly selected value from this domain. In this sense the membership function indicates the degree of possibility that a particular item is a member of a fuzzy set A.

In many practical situations the realistic modeling of phenomena necessitates the use of irregularly shaped, nonlinear membership functions, concave or convex, continuous or discontinuous, especially as a designer intends to obtain a better approximation of the model variables. However, the enumerated previously membership functions, particularly linear, triangular or trapezoidal are commonly used in fuzzy set applications, mainly for reasons of computational simplicity. The linear membership functions are also used in the mathematical formulations of optimization problems in a fuzzy environment.

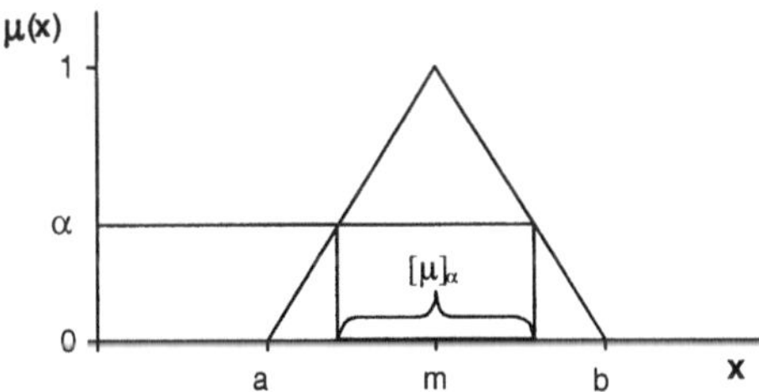

Figure 5. Definition of the α-cuts

2.4 Alpha-cuts

It should be pointed out that there is no unique fuzzy set representation by a membership function. Taking into account the possible applicability of fuzzy set concept the so-called horizontal representation of fuzzy sets is introduced by using their alpha cuts (α-cuts) instead of the membership functions $\mu(x)$, called vertical representation.

Let $\mu \in F(x)$ and $\alpha \in [0, 1]$. The set

$$[\mu]_\alpha = \{ x \in X \mid \mu(x) \geq \alpha \} \tag{15}$$

is called an α-cut of the membership function μ(x).

Let μ be the triangular function on IR given in Fig. 4a . The α-cuts of μ are in this case defined as follows:

$$[\mu]_\alpha = \begin{cases} [a + \alpha \cdot (m - a), b - a \cdot (b - m)] & \text{if } 0 < \alpha \leq 1 \\ IR & \text{if } \alpha = 0 \end{cases} \tag{16}$$

2.5 The Fuzzy Knowledge Base – Fuzzy Relations

Fuzzy logic concepts pertinent to the present work are outlined herein. Detailed reviews of the subject can be found in the literature [13,14]. Variability (fuzziness-uncertainty in a material data) is introduced by specifying a membership function (e.g. a normalized, in the sense of eq. (13), possibility distribution) denoted by μ. Let x_i (i=1,2,...N) means an input data vector having N independent components and y is an output data (an response quantity – one component). Each of the components (x_i ; y) is associated with an arbitrarily selected form of the membership functions ($\mu(x_i)$; $\mu(y)$). They may be different for each of the components (x_i ; y) – Fig.6.

Let us note also that both the number and the form of the membership functions are strongly dependent on the prescribed interval of the data variations, i.e. $[x_1^L, x_1^R]$ in Fig. 6. The dimensions of the input and the output spaces are defined by the intervals of the variables, where the superscripts L and R correspond to the left most (minimum) and the right most values (maximum), respectively. In order to build on appropriate Fuzzy Knowledge Base the whole process is divided into the following steps:

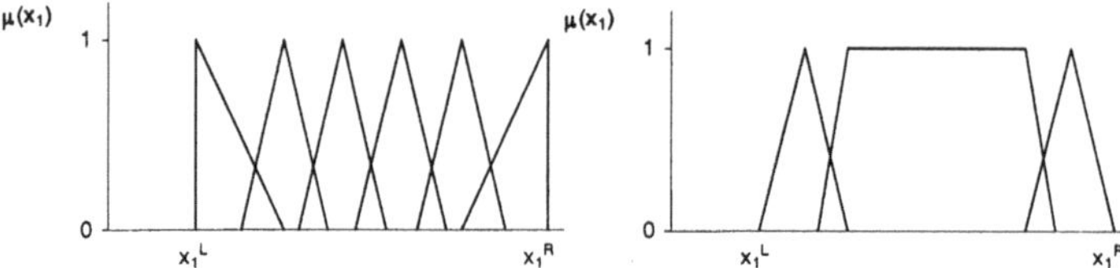

Figure 6. Possible selection of the membership functions $\mu(x_1)$

Divisions of the input and output intervals into $2N_{x_i}+1$ ($2N_y+1$, respectively) regions – see Fig. 7. Each of the regions is directly associated with the definitions of the appropriate membership function $\mu^J(x_i)$, J=1,2,.... $2N_{x_i}+1$

Fuzzy rules definitions. After the above divisions the details of the knowledge base design may be done. The procedure is based on the experimental data, for instance, if x_1 corresponds to the value of the stress level and x_2 to the number of cycles, we can measure the values of the Young modulus equivalent in our approach to the variable y. Thus, from the experiment we have a set of experimental functions describing the relation (in general it is non-linear)

$$y = f_i\,(x_1,x_2),\quad i=1,\dots,N \tag{17}$$

where the subscript i denotes the number of the experimental results and $f_i \neq f_j$. Of course, from the experiment the relation (17) is not a continuous one but it is given in a discretized form. If we have the set of experimental data $(\bar{x}_i, \bar{y})$ using the membership functions plotted in Fig. 7 one can derive a unique relation between the data, according to the following rule:

$$\bar{x}_1 \in A_3 \quad \text{if} \quad \mu^3(\bar{x}_1) = Max[\mu^J(\bar{x}_1)], \quad J=1,...,5 \tag{18}$$

Finally, we can easily formulate the set of relations in the following form:

$$\text{if} \quad \bar{x}_1 \in A_3 \quad \text{and} \quad \bar{x}_2 \in B_1 \quad \text{then} \quad \bar{y} \in C_2 \quad \text{etc.} \tag{19}$$

Of course, it is possible to get from the experiments two different rules, such as e.g.:

$R^{(1)}$: if $x_1^{(1)} \in A_1$ and $x_2^{(1)} \in B_1$ $y^{(1)} \in C_1$

$R^{(2)}$: if $x_1^{(2)} \in A_1$ and $x_2^{(2)} \in B_1$ $y^{(2)} \in C_2$

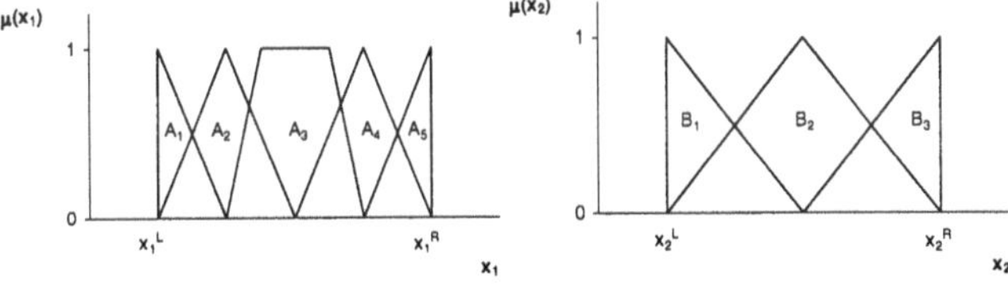

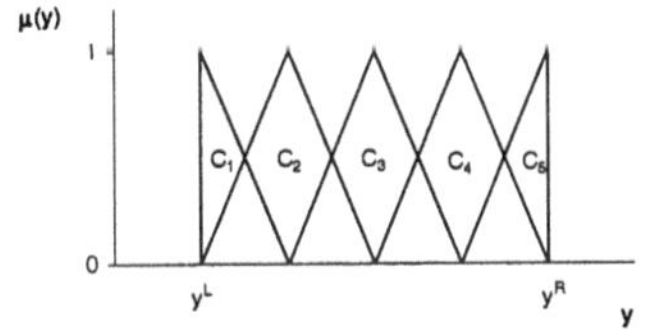

Figure 7. Possible definitions of the input and output membership functions

Their correctness is verified by the computation of the value

$$\mu(R^{(1)}) = \mu^1(x_1^{(1)}) \cdot \mu^1(x_2^{(1)}) \cdot \mu^1(y^{(1)}) \tag{20}$$

If $\mu(R^{(1)})$ is greater then $\mu(R^{(2)})$ then the first rule is attached to our Fuzzy Knowledge Base.

As it may be noticed the set of experimental data associated with the assumed form of the membership functions allows us to build the table shown in Table 1.

Relation (17) is strictly determined but only in the form of boxes presented in Table 1. It is necessary to define the continuous relation between input and output variables. Let us introduce the degree of the activity of the k-th rule in the following form:

$$\tau^{(k)} = \mu^{J_1(k)}(x_1) \cdot \mu^{J_2(k)}(x_2) \tag{21}$$

and using the center average defuzzification method one can find:

$$y = \frac{\sum_{k=1}^{N} \tau^{(k)} \cdot y^{(k)}}{\sum_{k=1}^{N} \tau^{(k)}} \tag{22}$$

Table 1 Fuzzy Knowledge Base (see eq. (19)).

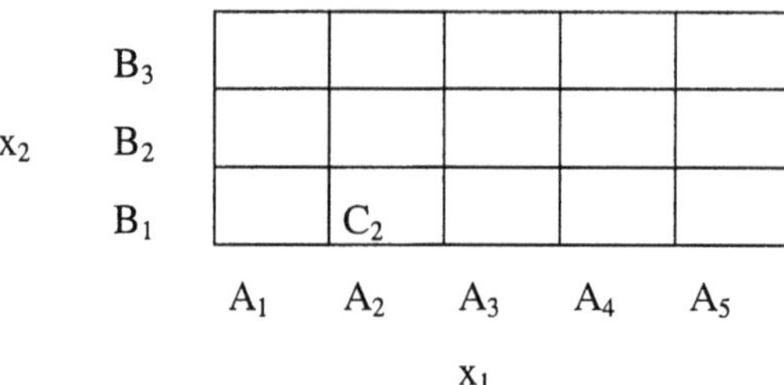

x_2		A_1	A_2	A_3	A_4	A_5
	B_3					
	B_2					
	B_1		C_2			

x_1

Of course, the above presented method may be easily extended for an arbitrary number of input and output data if the required experimental data are available – see also Section 7.

2.6 Fuzzy Set Probabilities

Having information about the membership representation of the value x_1 (see e.g. Figs 7, 8) it is possible to analytically express the event x_1 belongs to the fuzzy set A_3. Zadeh [17] has proposed to define the probability of the above event in the following way:

$$P(x_1 \in A_3) = \sum_i P(x_1^{(i)}) \cdot \mu_{A_3}(x_1^{(i)}) \tag{23}$$

where $x_1^{(i)}$ denotes experimental data drawn in the form of the membership function $\mu_{A_3}(x_3)$. $P(x_1^{(i)})$ denotes the classical probability of the events $x_1^{(i)}$ and for instance is equal to 0.1. If $\mu_{A_3}(x_1^{(i)})=1$ for each of the experimental data then probability described by eq. (23) is identical to the classical one, i.e. 0.3. In the opposite case as the membership function $\mu_{A_3}(x_1)$ has been built on the basis of three experimental data and for each of them the values of the membership function are equal e.g. 0.1 , 1 , 0.7 then $P(x_1 \in A_3)=0.18$ - eq. (23). Using the above definition of the fuzzy set probabilities one can obtain the probabilities of the event that the experimental results belong to the assumed interval, denoted in the analyzed example as the set A_3.

3 The Vertex Method – Computational Analysis

Let us introduce N fuzzy parameters describing material or geometric parameters of a composite structure considered. They membership functions are discretized using several α-cuts – eq. (16) . Considering the left and right end points of the α-cuts intervals $[\mu]_\alpha$ (see Fig. 7) for all fuzzy parameters one can find the total number of the combinations $N_{c/\alpha}$ per α-cut in the following form:

$$N_{c/\alpha} = \begin{cases} 2^N & for\ 0 \le \alpha < 1 \\ 1 & \text{for } \alpha = 1 \end{cases} \tag{24}$$

An output response denoted by p is an unknown function of the input fuzzy parameters x_i (i=1, 2, ..., N), so that:

$$p = f(x_1, \ldots, x_N) \tag{25}$$

Using the α-cut concept combined with the binary representation (24) of the fuzzy parameters x_i (i=1, 2, ..., N) the relation (25) can be rewritten in the abbreviated form:

$$p = f(C_{\alpha,j}),\ j = 1, 2, \ldots, N_{c/\alpha} \tag{26}$$

Since the output response p as a function of fuzzy parameters is a fuzzy set the corresponding interval in p is obtained from the relation [38]:

$$[\, p_\alpha^L, p_\alpha^R \,] = [\, \min_{\lambda,j} f(C_{\lambda,j}), \max_{\lambda,j} f(C_{\lambda,j});\ \lambda \ge \alpha,\ j = 1, 2, \ldots, N_{c/\alpha} \tag{27}$$

It may be seen that the relation (27) allows one to obtain a scatter of output parameters and then build the appropriate probability distributions and reliability functions by a sweep of α-cuts at different possibility levels.

In order to conduct the computations and to evaluate the upper and lower bounds of the output response (27) it is necessary to identify the deterministic method of the definition of the function f given in eq. (25). It can be defined in a purely analytical way or alternatively in a purely numerical way. Using the relation (27) it is possible to analyze the synergistic effects of fuzzy input parameters. On the other hand, the output response can be also understood in the sense of the fuzzy relation and the acceptance (or not) of the output value is derived with the use of any Fuzzy Knowledge Base – see Section 2.5, of course, under the condition that such a base exists.

The function f existing in eq. (25) may describe an arbitrary failure criterion for composites, e. g. buckling, delamination, the first-ply-failure etc., whereas the symbol p denotes the corresponding value of the failure load.

4 Mechanical Properties of Composite Materials

4.1 Micromechanics Model of Unidirectional Composites

The Aboudi micromechanics model [39] is included to study the effects of micromechanics constituent properties as fuzzy variables.

The development of the model of repeating cells is based on the assumption that the unidirectional composite is represented by fibers which are aligned in the x_1–direction and distributed regularly in the matrix and finally the components form a doubly periodic array in the x_2 and x_3 directions – see Fig.9. The fibers have a rectangular cross-section (h_1, l_1) and are arranged at distances h_2, l_2 apart. The cell is further divided into four subcells, each of them with a local coordinate system. The subcells are numbered using the pair of Greek indices α, β=1,2 – see Fig.9. The micromechanical analysis is performed on this representative cell by the first-order expansion of the displacement field in each subcell. By imposing the continuity conditions at the boundaries of the subcells in the average sense closed-form solutions for the average stress in the cell in terms of the average strain can be obtained. Without listing the details the micromechanics analysis leads to the following overall stress-strain relationship:

$$[\bar{\sigma}] = [Q][\bar{\varepsilon}] \tag{28}$$

where $\bar{\sigma}, \bar{\varepsilon}$ denote the average stresses and strains, respectively, whereas the Q_{ij} are the effective elastic constants of the unidirectional composite.

The material constants Q_{ij} are given in the functional form as:

$$Q_{ij} = f(C_{ij}^{(\alpha\beta)}, h_\alpha, l_\beta) \tag{29}$$

and $C_{ij}^{(\alpha\beta)}$ represent the constitutive properties of the (α,β) subcell, i.e. fibers or matrix in the assumed transversely isotropic material model of fiber reinforced composites.

At the micromechanics level, the transversely isotropic material is expressed in terms of the five engineering constants. They are functions of the material variables, i.e. of the fiber and the matrix properties: $E_{f11}, E_{f22}, G_{f12}, \nu_{f12}, \nu_{f13}$, E_m, ν_m, V_f, where f denotes fiber property, m – a matrix property, and V_f is the fiber volume ratio. For simplicity the superscripts (α,β) are omitted.

The impreciseness in the modeling is obvious: the ideal forms of the constituents and of representative cells, an influence of the interface is completely eliminated – there are two components in the model fibers and matrices, the matrix and fibers have constant values (there is no defects inside them) etc. In fact, the comparison of theoretical predictions with experimental data (see Refs [40, 41]) gave some discrepancies in the results.

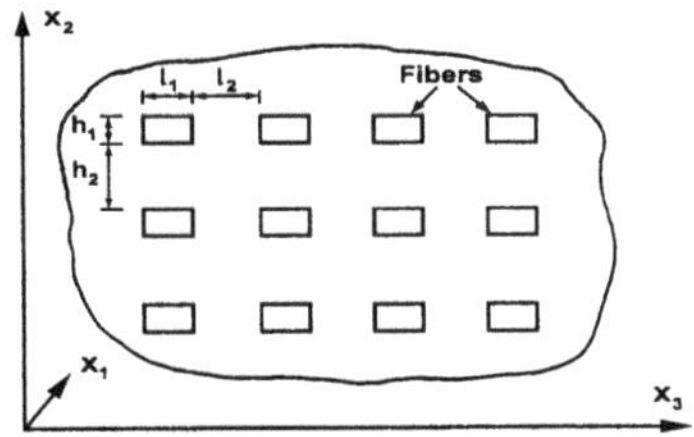

a) double periodic array of rectangular fibers extending in the x_1 direction

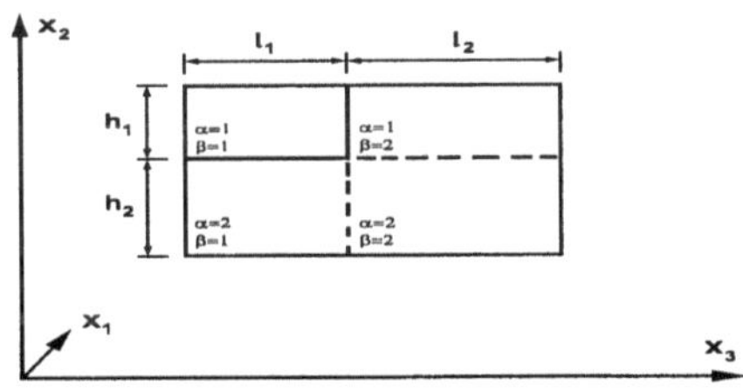

b) representative cell

Figure 8. Micromechanics subcell geometry

Now, let us assume that the variables in eq. (27) characterizing the fiber and matrix properties, the fiber volume fraction and the representative cell geometry are fuzzy variables and their fuzziness is described by a triangular membership function – eq. (6) and Fig. 4a. Since in the mechanical sense the problem is reduced to 2-D planar problem the fiber volume fraction can be derived from the geometrical relations (Fig. 8) as follows:

$$V_f = \frac{h_1 l_1}{(h_1 + h_2)(l_1 + l_2)} \tag{30}$$

It is assumed that the geometrical ratios defining the cell geometry may vary according their membership functions but in such a way that the fiber volume fraction (30) is constant, i.e. the total field of each cell is constant. The variability (understood as fuzziness) of material and geometrical parameters is taken as equal to ±10 %. The nominal (average) values correspond to α=1 (see Fig. 5) whereas the parameters of the triangular membership function given by eq. (16) are following :

$$m = \frac{a+b}{2}, \quad a = 0.9 \cdot m \quad b = 1.1 \cdot m \tag{31}$$

Having the analytical form of the relation (25) (see eq. (29)) and using the procedure proposed in the section 3 for a given α -cut one can evaluate the upper (the right end point of the interval (27)) and lower bounds (the left end point of the interval (27)) of the effective transverse Young modulus. The numerical results are evaluated for the glass/epoxy, i.e. the identical material as discussed in Ref [40]. It is interesting to note that the presented in Ref [41] comparison of the theoretical and experimental data shows that the experimental values are located rather above the nominal curve (the solid line in Fig. 9) what directly corresponds to the obtained results. The lower and upper bounds plotted in Fig. 9 (denoted by the superscripts L and R), and evaluated with the use of eq. (27) are just not symmetrical with respect to the nominal curve (α =1)

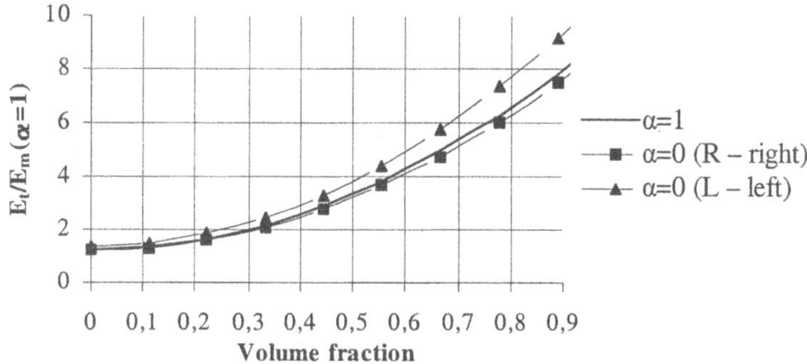

Figure 9. Distributions of the effective transverse Young modulus at $\alpha = 0$ and 1 for unidirectional composites

4.2 Fuzziness in Textile Composites

The development of innovative fiber architecture and textile manufacturing technology has significantly expanded the potential of fiber reinforced composites. An emerging area is textile composites reinforced with 3-D preforms such as woven, braided and knitted fabrics. The integrated fiber network provides stiffness and strength in the thickness direction thus reducing the potential for the inter-laminar failure.

However, on the other hand, more complicated than for unidirectional composites geometrical construction of textile preforms involves even higher degree of uncertainty (imprecision, fuzziness) in the evaluation of their elastic properties, particularly due to much more complicated geometrical parameters describing the constituents. The macroscopic stiffness of 3-D textile composites can be obtained from the macro-cell definition using both 2-D planar idealization of fibers architecture – Fig. 10 or 3-D – Fig. 11.

To demonstrate the impreciseness in the modeling the elastic constants of twisted yarn composites are predicted based upon the unit cell structure and volume averaging of elastic constants of constituent materials – see Ref [42]. Similarly as previously the composite cylinder is assumed to have transversely isotropic properties.

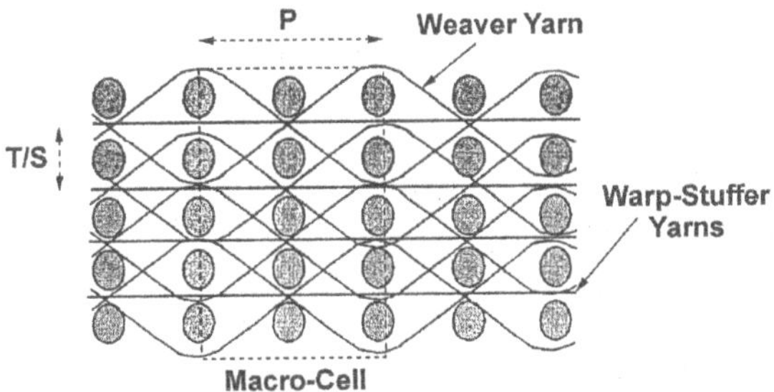

Figure 10. Idealized microstructure of a 3-D layer-to-layer angle interlock weave macro-cell – a 2-D cross-section (the warp-weaver yarn direction vs. the thickness direction)

The geometry of twisted yarns is characterized by the twist angle θ and the number of yarns twisted. Denoting the pitch length as h and the diameter of a twisted yarn as D the twist angle can be expressed in the following way:

$$tan\,\theta = \frac{\pi D}{h} \tag{32}$$

The diameter of the individual yarn is expressed as:

$$d = \sqrt{\frac{4\lambda}{\pi\kappa\rho n}} \tag{33}$$

where, λ – linear density of a twisted yarn, κ – fiber packing fraction, ρ – fiber density, n – number of yarns in the twisted yarn.

The fiber packing fraction is treated equivalently as the fiber volume fraction in the yarn. Assuming that the cross-section of an individual yarn is circular and the yarns are in contact with one another the diameter of the twisted yarn can be obtained from the following geometrical relation:

$$D = r_n d \tag{34}$$

where from the geometrical relations r_n has the following values:

$$r_2 = 2, \quad r_3 = 1 + \sqrt{2}, \quad r_4 = 1 + 2/\sqrt{3} \tag{35}$$

Thus, from eqs. (32) – (34) the fiber packing fraction is obtained as:

$$\kappa = \left(\frac{r_n}{h tan\theta} \right)^2 \frac{4\pi\lambda}{\rho n} \tag{36}$$

To determine the elastic properties of twisted yarns it is sufficient to consider one yarn in a length of one turn, regardless of the number of yarn bundles. The mechanical properties are predicted from yarn properties by averaging the stiffness (compliance) of each yarn through coordinate transformation. Let an infinitesimal yarn segment is defined in the global coordinate system (x, y, z) by the circumferential angles φ and $d\varphi$ and fibers are assumed to align along the axial direction corresponding to the 3-axis (perpendicular to the x-axis) in the local coordinate system (1,2,3) – see Fig. 12. Using the classical transformation rules the compliance matrix of the unidirectional composite cylinder [S] in the local coordinate system (1,2,3) is transformed to the matrix [S'] in the (x, y, z) coordinate system:

$$[S'] = [T]^{Tr} [S][T] \tag{37}$$

where the transformation matrix [T] takes the following form:

$$[T] = \begin{pmatrix} \alpha_1^2 & \alpha_2^2 & \alpha_3^2 & 2\alpha_2\alpha_3 & 2\alpha_1\alpha_3 & 2\alpha_2\alpha_1 \\ \beta_1^2 & \beta_2^2 & \beta_3^2 & 2\beta_2\beta_3 & 2\beta_1\beta_3 & 2\beta_2\beta_1 \\ 0 & \gamma_2^2 & \gamma_3^2 & 2\gamma_2\gamma_3 & 0 & 0 \\ 0 & \beta_2\gamma_2 & \beta_3\gamma_3 & \beta_2\gamma_3 + \beta_3\gamma_2 & \beta_1\gamma_3 & \beta_1\gamma_2 \\ 0 & \gamma_2\alpha_2 & \gamma_3\alpha_3 & \gamma_2\alpha_3 + \gamma_3\alpha_2 & \gamma_3\alpha_1 & \gamma_2\alpha_1 \\ \alpha_1\beta_1 & \alpha_2\beta_2 & \alpha_3\beta_3 & \alpha_2\beta_3 + \alpha_3\beta_2 & \alpha_1\beta_3 + \alpha_3\beta_1 & \alpha_1\beta_2 + \alpha_2\beta_1 \end{pmatrix} \tag{38}$$

and:

$$\begin{aligned} &\alpha_1 = cos\,\theta, \quad \alpha_2 = -sin\,\theta\, sin\,\varphi, \quad \alpha_3 = sin\,\theta\, cos\,\varphi, \\ &\beta_1 = sin\,\theta, \quad \beta_2 = cos\,\theta\, sin\,\varphi, \quad \beta_3 = -cos\,\theta\, cos\,\varphi, \\ &\gamma_1 = 0, \quad \gamma_2 = cos\,\varphi, \quad \gamma_3 = sin\,\varphi, \end{aligned} \tag{39}$$

Finally, the compliances [S'] are averaged through one turn of the twist since it is assumed that the composite yarn consists of the infinitesimal segments in series, thus:

$$S_{ij}^{av} = \frac{1}{2\pi} \int_0^{2\pi} S'_{ij} d\varphi, \quad i, j = 1,...,6 \tag{40}$$

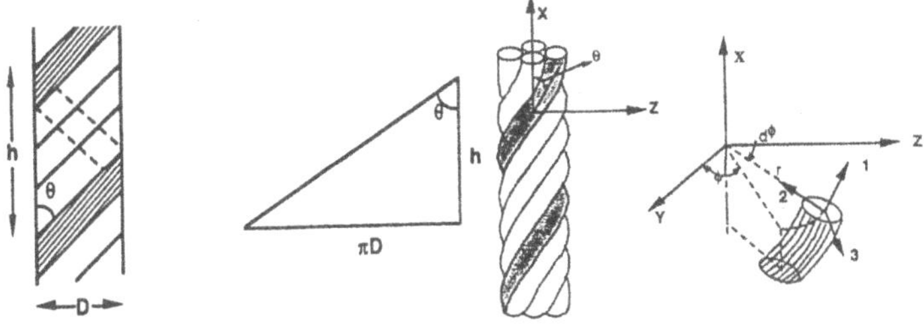

Figure 11. The idealized geometry of twisted yarn composites

Using the methodology presented in Section 3 and the same variability of parameters described in Section 4.1 the upper and lower bounds of the Young modulus have been evaluated. The twist angle θ, λ – linear density of a twisted yarn and the Young's moduli of fibers and matrix have been treated herein as the fuzzy parameters. However, the fiber volume fraction is assumed to be constant as the geometrical parameters of the twisted yarn vary. The nominal properties, both material and geometrical, are identical to those prescribed in Ref [42]. The results of computations are plotted in Fig.13. As it may be seen, again there is no symmetry with respect to the nominal curve denoted by the value α =1 (the solid line). Similarly, as in the case of the unidirectional fibers the experimental data lie above the nominal curve corresponding to the theoretical predictions. However, in this case there is only one experimental value given in Ref [42] and in this way it is rather difficult to verify the correctness of the used fuzzy approach.

At the end of this section it is worth mentioning that the presented method allows one to build the appropriate membership functions for the evaluated values of the Young's moduli since the plots shown in Figs. 9 and 12 can be supplemented by the curves corresponding to other values of the α parameter belonging to the interval [0,1] – compare with Fig.6 and eqs. (15), (16).

5 Fuzzy Set Analysis of Limit Load Carrying Capacity

Damage of composite materials is one of the most interesting problems in view of engineering applications. However, the scatter of the experimental data is even higher then in the modeling of mechanical properties since much more factors can affect failure loads, not saying about the variety of failure modes. This is the reason that for composite structures the concept of the limit load carrying capacity (LLCC) has been introduced by Muc *et al.* [43]. Briefly speaking the fundamental idea of the LLCC is based on the evaluation of the lower bound envelopes of different failure loads corresponding to the analyzed composite structure subjected to the prescribed loading and boundary conditions and having uniquely defined laminate topology, here understood in the sense of the deterministic, nominal values of all parameters. By different failure loads we mean loads corresponding to various failure modes encountered in the analysis of composite structures, i.e. delaminations, matrix cracking (the First-Ply-Failure – FPF), global or local buckling, fibers disbonding etc. In details, in the present section three types of failure modes are considered: delaminations, FPF and global buckling of structures. The results of the analysis are illustrated on the example of compressed rectangular laminated multilayered plate.

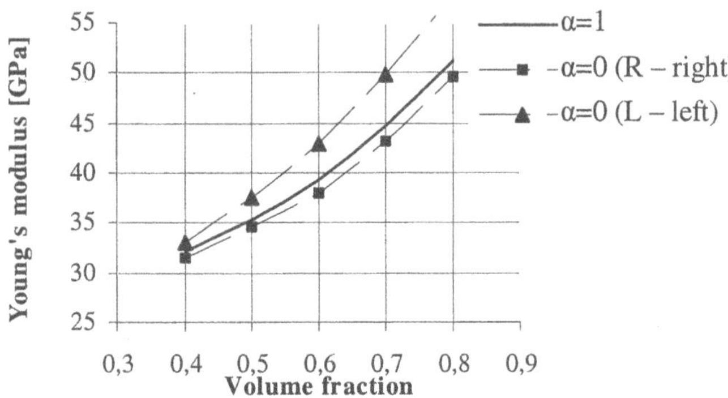

Figure 12. Variations of Young's modulus with fiber volume fraction for twisted yarns with n=2

It should be emphasized that the current analysis is limited to the evaluation of failure loads with the help of the appropriate global failure criteria. It is also possible to use a micromechanical approach but those problems are not discussed herein. The description of methods and criteria introduced and used in micromechanics of composite materials in a fuzzy environment are presented e.g. in Refs [44,45] – a problem of intra-laminar cracks.

5.1 Delaminations

Let us consider a buckling problem of a rectangular plate having an ellipsoidal crack located at the plate center and subjected to a cyclic compression (fatigue analysis). The total crack area is denoted by the symbol A. It is assumed that the crack area may vary depending on the loading (cyclic) conditions. Analysis of delamination initiation is carried out either by mechanics of materials approach [46,47] or a fracture mechanics approach [48] – [50]. Since the delamination growth can be attributed to a mixed mode of fracture, the criterion for the initiation of the delamination growth was selected as follows [51]:

$$RB = g_I^{\delta} + g_2^{\gamma}, \quad g_I = \frac{G_I}{G_{Ic}}, \quad g_2 = \frac{G_{II}}{G_{IIc}} \tag{41}$$

where *RB* is the failure index, G_{Ic} and G_{IIc} are the critical strain energy release rates corresponding to mode I and II fracture, respectively. δ and γ are coefficients – it was found that $\delta = \gamma = 1$ provides the best fit to the experimental results. A delamination would start to propagate when $RB \geq 1$.

A comparison of fatigue tests with different stress ratios R indicates that fatigue damage manifests itself most severely at R=-1 (see e.g. Ref [52]) and this value was taken in the analysis. In addition, it has been observed that under the stress ratio R=-1 at comparatively low stress amplitudes matrix cracks develop rather early, and then at increasing cycle numbers precipitate localized delaminations. In order to take into account ply degradation due to matrix cracking it is assumed the simplest degradation model of the material properties describing plies oriented at 90°, i.e.: $E_1 = \beta E_1$, $E_2 = G_{44} = 0$, where β is a degradation factor. The more complicated model considering fuzzy set approach to the analysis of intra-laminar crack propagation is discussed in Ref. [45] but now we restrict our investigations to the inter-laminar crack behavior only.

In the numerical analysis five parameters are considered to be fuzzy, namely, the degradation factor β and four parameters describing the failure criterion (eq. (41), i.e. δ, γ , and G_{Ic}, G_{IIc}. Russell and Street [53] measured fracture toughness of common composites and obtained the scatter in experimental data equal to ±5 %. In our analysis the variability (understood as the fuzziness) of these parameters is taken to be equal to ±10 %. The identical variability is assumed for the rest of parameters, i.e. β, δ and γ. Their nominal values corresponding to α =1 (Fig. 5) are taken as follows: β = 0.6, γ =1.0, δ =1.0.

The vertex of the triangular membership functions is always located at the middle of the intervals describing variations of the parameters. It should be emphasized that the form of the membership function is a hypothesis and the triangle plotted in Fig.6 can be replaced by other function satisfying experimental conditions and results or simply the assumptions of the theoretical modeling. It is assumed that the analyzed herein plate is made of the carbon/epoxy and their static material properties are following: $E_1 =$ 134 GPa, $E_2 = E_3 =$ 10.2 GPa, $G_{12} = G_{13} =$ 5.52 GPa, $G_{23} =$ 3.43 GPa, $\nu_{12} = \nu_{13} =$ 0.3, $\nu_{23} =$ 0.49.

The different binary combinations $C_{\alpha,j}$ formed by failure coefficients δ, γ , and G_{Ic}, G_{IIc} result in different failure envelopes. For the nominal values of the above parameters the dimension-less failure curve is a straight line (α = 1). The failure envelopes being a lower and an upper bounds at α = 0 are not straight lines and in addition they are not symmetric with the respect to the nominal value at α = 1. The presented example of the sensitivity study demonstrates the synergistic effects of different fuzzy parameters. Although relatively small ranges of the variability in the input failure coefficients were selected, ± 10%, the resulting variability in the output failure envelopes is quite large. However, the symmetry or not of the upper and the lower bounds at α = 0 with respect to the nominal curve (α = 1) depends completely on the type of the fuzzy parameter. As it may be seen in Fig.14 the abovementioned curves describing lower and upper bounds are symmetric if we analyze the variability (fuzziness) of the degradation factor β parameter. It should be emphasized that such behavior is possible only as the buckling mode is identical for the whole ranges of parameters considered in the numerical analysis. Since it was investigated that the buckling mode for the compressed plate had the symmetric shape it was possible to obtain symmetry in the plot drawn in Fig.13.

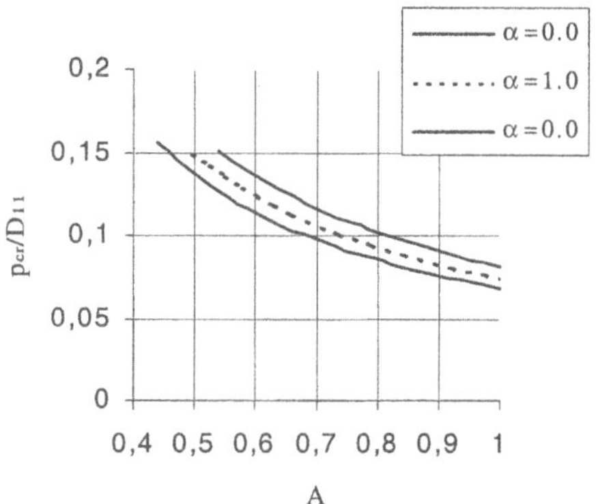

Figure 13. Dimensionless buckling loads of a rectangular plate with ellipsoidal crack.

5.2 First Ply Failure

The classical first ply failure (FPF) envelopes (e.g. the Tsai-Wu criterion) are described with the aid of a variety of material constants characterizing different modes of failures (tension, compression etc.). Those values may be treated as fuzzy ones and in this way the fuzzy set theory allows us to build the lower and upper bounds of the failure envelopes for each fiber orientations of plies in the laminate. Assuming that five strength constants are fuzzy parameters such envelopes have been built and the plot is presented in Fig. 14. For compressive deformations the uncertainty of failure locations (at plies having fibers oriented at 0^0 or 45^0) is the characteristic feature of the resulting curves.

The results clearly demonstrate that the classical deterministic approach to the FPF analysis can lead to inconsistency and incorrectness with experimental data. For compression in both directions it is possible to experimentally obtain the FPF initiation at plies oriented both at 0^0 and 45^0. They simply show the limitations of the classical global elastic approach since the matrix cracking associated with the FPF is connected with inelastic deformations (in the global sense), especially for fibers oriented at 45^0 or with the appearance of intra-laminar cracks (a micromechanical approach).

5.3 Global Buckling

To evaluate global buckling loads numerical studies are performed on compressed square plate. The numerical FE analysis is carried out in the elastic geometrically linear range only, with the use of the four node quadrilateral shell elements (NKTP 32) employing the first order transverse shear deformation plate/shell theory. The geometric and material characteristics are following: $E_x = 280$ GPa, $E_y = 12$ GPa, $G_{xy} = 7$GPa,$G_{xz} = 0.6G_{xy}$,$G_{yz} = G_{xy}$,$\nu_{xy}= 0.28$, $t/a = 0.1$, where t is a plate thickness and a is a plate length, respectively. Four parameters have been considered as fuzzy variables: the total thickness t, the Young moduli E_x, E_y and the Kirchhoff modulus G_{xy}.

Similar to the previous results, in the present study it is assumed that the membership functions of the fuzzy parameters are triangular as shown schematically in Fig. 5 (see also eq. (16)). Again the variability (fuzziness) is taken to be equal ±10%, which falls within typical ranges of scatter in experimental data for static tests.

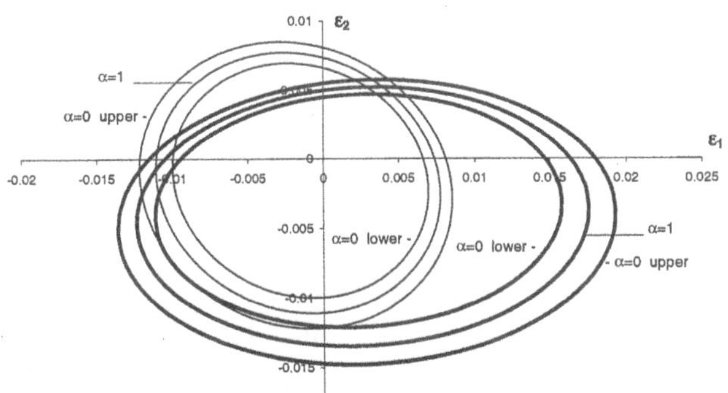

Figure 14. Upper and lower bounds of FPF envelopes –Tsai-Wu criterion (0^0 ▬▬ , —— 45^0)

The distributions of buckling pressures versus the angle of fiber orientations at α=0, 0.5 and 1.0 are plotted in Fig. 15. The upper and lower bounds (α=0 and 0.5) of the curves drawn for α=1.0 are not symmetric. It is necessary to point out that the interval (27) is strongly dependent on the fiber orientations as well as on the wave number in buckling.

Using the identical methodology it is possible to consider delaminations, FPF and buckling problems (in the conjunction with the FE analysis) for arbitrary types of laminated composite structures having different types of material and geometric imperfections or even analyze the whole technological process (e.g. the RTM process) including heating and curing of a resin.

6 Optimization Problems in a Fuzzy Environment

6.1 Formulation of the Problem

In general, the problem of the optimum design of structures (or materials what becomes also an objective in the composite materials area) depends on the definition of: (i) an objective function (functional), (ii) a set of inequality (or equality) constraints, (iii) design variables, and (iv) optimization algorithm. It is obvious that the first three factors may be classified into three different groups, i.e. deterministic (crisp), probabilistic (random) or fuzzy. The optimum design based on the deterministic or probabilistic approaches are rather well developed and known both from the theoretical, mathematical background as well as from numerical point of view.

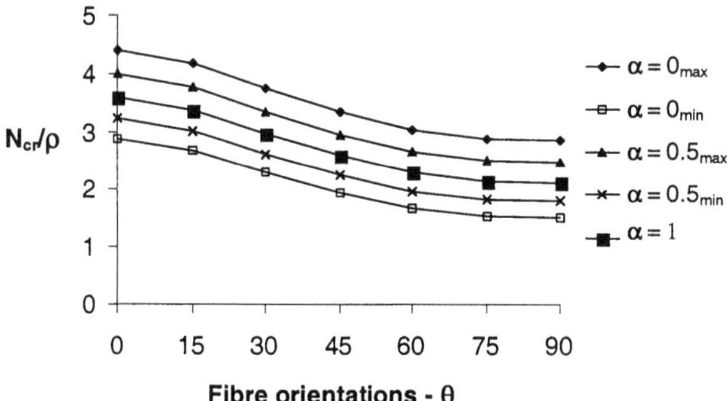

Figure 15. Distributions of buckling loads for compressed angle-ply plates

The optimum design in a fuzzy environment is a new approach although some attempts have been made in this area starting from the pioneering works of Zimmermann [30], Rao [54,55], Verdegay [31], Yazenin [32], Yeh & Hsu [33]. An application of fuzzy sets to optimization problems of structures made of composite materials have been given by Morton & Weber [34], Adali [35]. For a fuzzy optimization problem the fuzzy objective function and constraints are characterized by their membership functions.

In Table 2 there are compared two optimization problems deterministic and fuzzy. Of course, the minimum design problem may be simply reformulated to the maximum problem. As it may be noticed the definition of the constraints seems to be the difference in both formulations. However, as it will be demonstrated below the significant difference lies in the optimization algorithms and the reformulation of the fuzzy optimum design problems, since in a fuzzy environment searching for the optimum solutions is carried out not for the values of design variables s (a crisp problem) but for the membership functions.

Table 2 Comparison of the deterministic and fuzzy optimization problems

	Deterministic (crisp) formulation	Fuzzy formulation
Objective	Find: Min f(s)	Find: Min f(s)
Constraints	$g_j(s) \le b_j$, j = 1,2,...,m	$g_j(s) \in G_j$ (fuzzy sets), j=1,2,...m

s – design variables (a vector)

The fuzzy feasible region A is defined by considering all the constraints, i.e.:

$$A = \bigcap_{j=1}^{m} G_j \tag{42}$$

and the membership degree of any design variables s to a fuzzy feasible region A is given by the following condition :

$$\mu_A(s) = \underset{j}{Min}\{\mu_{g_j}(s)\} \tag{43}$$

This is the minimum degree of satisfaction of the design variables s to all the constraints. A design variables can be considered feasible if the objective function defines a non-empty fuzzy set D⊂A, the intersection of the membership functions of the objective and the constraints, such that:

$$D = \{\mu_f(s)\} \cap \left\{ \bigcap_{j=1}^{m} \mu_{g_j}(s) \right\} \tag{44}$$

From the fuzzy set the optimum solution $\bar{s}$ is selected as the design for which the membership is a maximum (see Fig.17), i.e.:

$$\mu_D(\bar{s}) = Max\mu_D(s) = Max[Min(\mu_f(s), \mu_{g_1}(s), ..., \mu_{g_m}(s))] \tag{45}$$

Thus, using the fuzzy formulation the initial minimum problem (see Table 2) is formulated as a MinMax problem. The latter problem is reformulated to the following maximum problem introducing an additional objective λ:

$$Find\ Max\ \lambda \tag{46}$$

subject to the constraints:

$$\lambda \le \mu_f(s), \quad \lambda \le \mu_{g_j}, j = 1,2,...,m \tag{47}$$

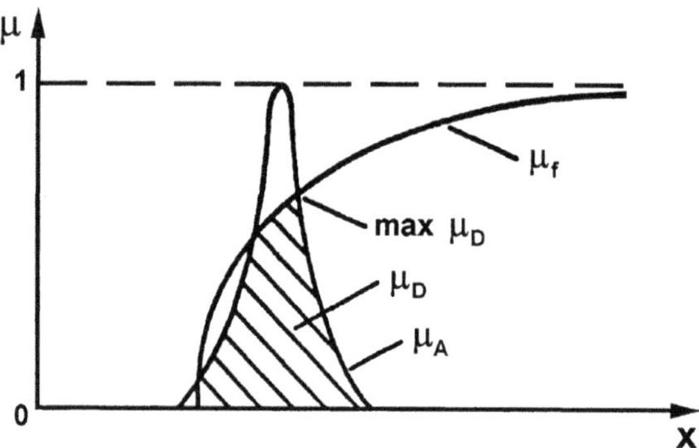

Figure 16. Optimal fuzzy decision-making

6.2 Illustrative Example

In order to demonstrate the sense of the definitions introduced in the previous section let us consider the following problem:

Find the optimal fiber orientations in an angle-ply rectangular axially compressed plate subjected to the buckling and strength constraints in the linear form (for details see Muc [56]). Using the formulation given by eqs. (45)-(47) and the linear qualifier membership functions (see Fig. 5h) the present optimization problem can be expressed in the following explicit form:

Find:

$$\underset{\theta}{Max}\,\alpha\ , \quad \alpha \in [0,1] \tag{48}$$

subject to the following constraints:

$$\alpha \le 1-(\sigma_1 - X_c)/(X_c^{max} - X_c); \qquad \alpha \le 1-(\sigma_2 - Y_c)/(Y_c^{max} - Y_c)$$

$$\alpha \le 1-(\sigma_{12} - S)/(S^{max} - S); \qquad \alpha \le 1-(\lambda - \lambda_{min})/(\lambda_{max} - \lambda_{min}) \tag{49}$$

where $\sigma_1, \sigma_2, \sigma_{12}$ denote the stress components, λ is a buckling load factor, and θ is a fiber orientation. Since from constraint conditions (49) it is possible to derive the explicit dependence on the fiber orientations the relations (49) can be replaced by the equivalent where the external load factor is a linear function of the α parameter and of trigonometric functions of the θ angle. In this way the optimization problem (48) can be solved in an analytical way. However, the problem (48) becomes more complicated if we have different fiber orientations in each of the plies constituting the laminate or the membership functions are not linear as assumed in the relations (49).

7 Analysis of the Experimental Data

Of course, experimental data represent always a various scatter of results and in this way the impreciseness (uncertainty) of the experimentally measured value that in our approach is treated as a fuzzy number. Usually, the experimental results are given in the form of the histogram. Using the histogram it is possible to propose a statistical distribution describing the data but with a margin of uncertainty since a form of statistical distributions is simply a hypothesis. Having the statistical distributions it is possible to construct the membership function for the measured value with the help of eq. (13). On the other hand it is possible also to construct the membership function with the aid of the histogram only. The procedure will be presented on the example of fatigue tests conducted by Lee *et al.* [57] for the UD specimens using the fuzzy set methodology. In that case the impreciseness in the evaluation of fatigue life is quite large and extends from 1,392 to 190,860 cycles. However, it should be emphasized that those numbers are typical in fatigue tests.

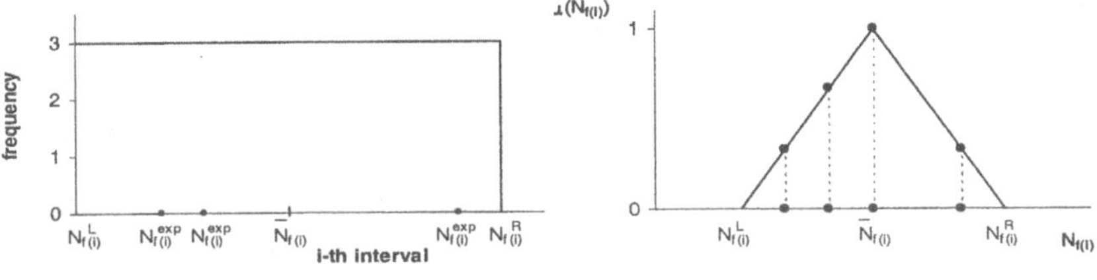

a) the histogram b) the membership function

Figure 17. Membership function representative of the experimental results

Each part of the histogram we intend to adjust to the membership function representation in the following manner (see also Fig. 17):

to find the mean value of the fatigue life $\overline{N}_{f(i)}$ in the ith interval to prescribe that the value of the membership function corresponding to the mean value of the fatigue lives is equal to 1

to assume that the values of the membership function at the ends (i.e. $N_{f(i)}^{L}$ and $N_{f(i)}^{L}$) of the i-th interval are equal to 0

to determine the value of the membership function to the given experimental data $N_{f(i)}^{exp}$

$$\mu(N_{f(i)}^{exp}) = \frac{N_{f(i)}^{R(L)} - N_{f(i)}^{exp}}{N_{f(i)}^{R(L)} - \overline{N}_{f(i)}} \tag{50}$$

With the use of the above-mentioned construction it is possible to build the Fuzzy Knowledge Base or probabilities for the given set of experimental data. The applicability of the fuzzy set method is presented in Fig. 18 being a plot of probabilities (defined by eq. (23)) of the fatigue-life experimental data discussed by Lee *et al.* [57]. It should be emphasized that we do not have a strict values of the fatigue-lives. They were estimated only from the histogram plotted in Ref. [57]. As it may be observed there are discontinuities of the curve plotted in Fig. 17 caused by the histogram form of the available experimental data.

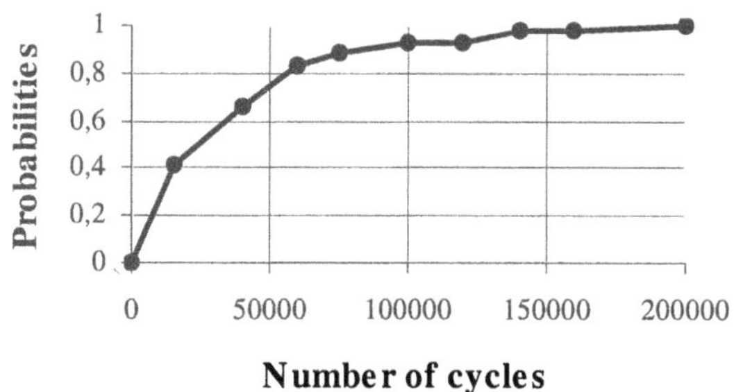

Figure 18. A fuzzy probability of experimental data.

An obvious advantage of a fuzzy-set based approach depends on the possibility of using all input and output parameters without introduction of any statistical distributions. However, in order to fit experimental results as much as possible to prescribed *a priori* membership functions (see Fig. 8) it is better to examine a few forms and study their influence on the final results.

A similar analysis can be conducted with the use of the Fuzzy Knowledge Base.

8 References

1. Taylor N.W. (1947), Mechanism of fracture of glass and similar brittle solids, J. Appl. Physics, 18, pp. 943-954.

2. Stuart D.A. and Anderson O.L. (1952), Dependence of ultimate strength of glass under constant load and temperature, ambient atmosphere, and time, J. Amer. Ceram. Soc., 36, pp. 416-424.

3. Charles R.J. (1958), Static fatigue of glass. I, Journal of Applied Physics, 29, pp. 1549-1553.

4. Charles R.J. (1958), Static fatigue of glass. II, Journal of Applied Physics, 29, pp. 1554-1560.

5. Mould R.E. and Southwick R.D. (1959), Strength and static fatigue of abraded glass under controlled ambient conditions, J. Amer. Ceram. Soc., 42, Part I, pp. 542-581, Part II, pp. 582-607.

6. Charles R.J. and Hillig W.B. (1961), The kinetics of glass failure by stress corrosion, published as part of the Symposium on the Mechanical Strength of Glass and Ways of Improving It, Union Scientifique Continentale du Verre.

7. Kies J.A. (1964), The strength of glass fibers and the failure of filament wound pressure vessels, NRL Report 6034, U. S. Naval Research Laboratory.

8. Schmitz G.K. and Metcalfe A.G. (1965), Characterization of flaws on glass fibers, Proc. 20th Anniversary Technical Conference, The Society of the Plastics Industry.

9. Schmitz G.K. and Metcalfe A.G. (1966), Stress-corrosion of E-glass fibers, I&EC Product Res. & Dev., 5, pp. 1-6.

10. Elishakoff I. (1983), Probabilistic methods in the theory of structures, John Wiley & Sons, New York.

11. Rao S.S. and Sawyer P. 1995), Fuzzy finite element approach for the analysis of imprecisely defined system, AIAA Journal, 33, no. 12, pp. 2364-2370.

12. Zadeh L.A. (1965), Fuzzy sets, Information and Control, 8, pp. 338-353.

13. McNeill F.M. and Thro E. (1996), Fuzzy logic: a practical approach, AP Professional, Fuzzy Logic CD-Rom Library, Electrographics Electronic Imaging, East Lansing, Michigan, USA.

14. Cox E. (1996), The fuzzy systems handbook, AP Professional, Fuzzy Logic CD-Rom Library, Electrographics Electronic Imaging, East Lansing, Michigan, USA.

15. Tsoukalas L.H. and Uhrig R.E. (1997), Fuzzy and neural approaches in engineering, A Wiley-Interscience Publication: John Wiley and Sons, New York, USA.

16. Dubois and Prade (1996), Fuzzy sets and system: theory and applications, AP Professional, Fuzzy Logic CD-Rom Library, Electrographics Electronic Imaging, East Lansing, Michigan, USA.

17. Zadeh L.A. (1996), Fuzzy sets and their application to cognitive and decision process, AP Professional, Fuzzy Logic CD-Rom Library, Electrographics Electronic Imaging, East Lansing, Michigan, USA.

18. Kosko B. (1991), Neural networks and fuzzy systems, Prentice Hall, Englewood Cliffs, NJ.

19. CD-ROM (1996), AP Professional, Fuzzy Logic CD-Rom Library, Electrographics Electronic Imaging, East Lansing, Michigan, USA.

20. Mathematica® (1995), Fuzzy logic pack, Wolfram Research, Inc., Champaign, Illinois.

21. Lallemand B., Plessis G., Tison T., and Level P. (1999), Neumann expansion for fuzzy finite element analysis, Engineering Computations, pp. 572-583.

22. Skrzypczyk J. and Burczynski T. (1997), The fuzzy boundary element method, Proc. XIII PCCMM, Poznań, pp. 1195-1202.

23. Abdel-Tawab K. and Noor A.K. (1999), A fuzzy-set analysis for a dynamic thermo-elasto-viscoplastic damage response, Computer and Structures, 70, pp. 91-107.

24. Muc A. and Kedziora P. (2001), A fuzzy-set approach to failure analysis of composite structures, Mech. Composite Mat., (in print).

25. Szeliga E. and Witkowski M. (1999), Fuzzy Monte Carlo Method for the stability assessment, Proc. XIV PCCMM, Rzeszów, pp. 353-354.

26. Brown C.B. and Yao J.T.P. (1983), Fuzzy sets and structural engineering, ASCE Journal of Structural Engineering, 109, pp. 1211-1225.

27. Valliappan S. and Pham T.D., Elasto-plastic finite element analysis with fuzzy parameters, International Journal for Numerical Methods in Engineering, 38, pp. 531-548.

28. Loskiewicz-Buczak A. and Uhrig R.E. (1993), Neural network – fuzzy logic diagnosis system for vibration monitoring, Proceedings of ANNIE-93, Artificial Neural Networks in Engineering Conference, St. Louis, MO.

29. Kohonen T. (1990), The self-organizing map, Proceedings of the IEEE, 78, no. 9.

30. Zimmermann H.J. (1976), Description and optimization of fuzzy systems, International Journal of General Systems, 2, no. 4, pp. 209-215.

31. Verdegay J.L. (1982), Fuzzy mathematical programming, Fuzzy Information and Design Process (Edited by M. M. Gupta and E. Sanchez), North-Holland, Amsterdam.

32. Yazenin A.V. (1987), Fuzzy and stochastic programming, Fuzzy Sets Syst., 22, pp. 171-180.

33. Yeh Y.C. and Hsu D.S. (1988), Structural optimization with uncertainty factors, Twelfth National Conference on Theoretical and Applied Mechanics, Taipei, Taiwan, R.O.C., pp. 565-571.

34. Morton S.K. and Webber J.P.H. (1990), Uncertainty reasoning applied to the assessment of composite materials for structural design, Engineering Optimization, 16, pp. 43-77.

35. Adali S. (1991), Fuzzy optimization of laminated cylindrical pressure vessels, Composite Structures, Elsevier Applied Science, London, pp. 249-260.

36. Norwich A.M. and Turksen I.B. (1984), A model for the measurement of membership and consequences of its empirical implementation, Fuzzy Sets and Systems, 12, no. 1, pp. 1-25.

37. Dombi J. (1993), Membership function as a evaluation, Fuzzy Sets and Systems, 35, no. 1, pp. 1-21.

38. Dong W. and Shah H.C. (1987), Vertex method for computing functions of fuzzy variables, Fuzzy Sets and Systems, 24, pp. 65-78.

39. Aboudi J. (1987), Closed form constitutive equations for metal matrix composites, International Journal Engineering Science, 25, no. 9, pp. 1229-1240.

40. Aboudi, J., (1989), Micromechanical analysis of composites by method of cells , Appl. Mech. Rev., Vol. 42, No 7, pp. 193-221. Aboudi, J., (1984), Effective behavior of inelastic fiber reinforced composites, International Journal Engineering Science, 22, pp.439-449.

41. Aboudi, J., (1982), A continuum theory for fiber reinforced elastic-viscoplastic composites, International Journal Engineering Science, 20, pp.605-621.

42. Byun J.H. and Chou T.W. (1995), Effect of yarn twist on the elastic property of composites, Proceedings of ICCM-10, Whistler, B. C., Canada, 4, pp. 293-299.

43. Muc A., Rys J., and Latas W. (1993), Limit load carrying capacity for spherical laminated shells under external pressure, Composite Structures, 25, pp. 295-303.

44. Muc A. and Kedziora P. (2000), A fuzzy set analysis for a fatigue damage response of composite materials, Proceedings of International Conference ICCST/3, Durban, pp. 417-422.

45. Muc A. and Kedziora P. (2001), A description of intra-laminar cracks with the use of fuzzy sets, Composite Structures, (in print).

46. Fish J.C. and Lee S.W. (1988), Tensile strength of tapered composite structures, Proceedings of the AIAA/ASME/ASCE/AHS/ASC 29th Structures, Structural Dynamics and Materials Conference, Part I, AIAA Paper 88-2252.

47. Curry J.M., Johnson E.R., and Starnes J.H. (1992), Effect of dropped plies on the strength of graphite/epoxy laminates, AIAA J., 30, pp. 82-88.

48. Winsom M.R. (1991), Delamination in tapered unidirectional glass/epoxy under static tension loading, Proceedings of the AIAA/ASME/ASCE/AHS/ ASC 32nd Structures, Structural Dynamics and Materials Conference, Part II, AIAA Paper 91-1142

49. Salpeker S.A., Raju I.S., and O'Brien T.K. (1988), Strain energy release rate analysis of delamination in tapered laminates subjected to tension loading, Proceedings of the

American Society for Composites, 3rd Technical Conference, Technomic Publ., Lancaster.

50. Trethewey B.R., Gillespie J.W., and Wilkins D.J. (1990), Inter-laminar performance of tapered composite laminates, Proceedings of the American Society for Composites, 5th Technical Conference, Technomic Publ., Lancaster.

51. Kanninen M.F. and Popelar C.H. (1985), Advanced Fracture Mechanics, Oxford Engineering Science Series 15, Oxford University Press.

52. Bergmann H.W., Prinz R. (1989), Fatigue life estimation of graphite/epoxy laminates under consideration of delamination growth, Int. J. Num. Meth. Eng., 27, pp. 323-341.

53. Russel A.J. and Street K.J. (1987), The effect of matrix toughness on delamination static and fatigue fracture under mode II shear loading of graphite composites, Toughened Composites, ASTM STP 937, American Society for Testing and Materials, Philadelphia.

54. Rao S.S. (1987), Multi-objective optimization of fuzzy structural systems, International Journal for numerical Methods in Engineering, 24, pp. 1157-1171.

55. Rao S.S. (1992), Fuzzy goal programming approach for structural optimization, AIAA Journal, 30, no. 5, pp. 1425-1432.

56. Muc A. (1988), Optimal fiber orientations for simply -supported plates under compression, Composite Structures, 9, pp.161-172.

57. Lee J., Harris B., Almond D.P., and Hammett F. (1997), Fiber composites fatigue-life determination, Composites: Part A, 28A, pp. 5-15.

Chapter 3 Soft Computing and Density Functional Theory in the Design of Safe Textile Chemicals

Les Sztandera[1], Mendel Trachtman[2], Charles Bock[2], Janardhan Velga[3], Ashish Garg[3]

[1]Computer Science Department, [2]Chemistry Department, [3]School of Textiles and Materials Technology, Philadelphia University, Philadelphia, PA 19144, USA

Summary: This research focuses on the use of soft computing to aid in the development of novel, state-of-the-art, non-toxic dyes which are of commercial importance to the U.S. textile industry. Where appropriate, modern molecular orbital (MO) and density functional (DF) techniques are employed to establish the necessary databases of molecular properties to be used in conjunction with the neural network approach. In this research, we focused on: 1) using molecular modeling to establish databases of various molecular properties of azo dyes required as input for our neural network approach; 2) designing and implementing a neural network architecture suitable to process these databases; and 3) investigating combinations of molecular descriptors needed to predict various properties of the azo dyes.

Keywords: Fuzzy entropy, Feed-forward neural networks, Molecular modeling, Density Functional Theory

1 Introduction

This research involves the integration of fuzzy entropies (used in the context of measuring uncertainty and information) with computational neural networks. An algorithm for the creation and manipulation of fuzzy entropies, extracted by a neural network from a data set, is designed and implemented. The neural network is used to find patterns in terms of structural features and properties that correspond to a desired level of activity in various azo dyes. Each molecule is described by a set of structural features, a set of physical properties and the strength of some activity under consideration. After developing an appropriate set of input parameters, the neural network is trained with selected molecules, then a search is carried out for compounds that exhibit the desired level of activity. High level molecular orbital and density functional techniques are employed to establish databases of various molecular properties required by the neural network approach.

The structural and electronic properties of the positional isomers of monomethoxy-4-aminoazobenzene (n-OMe-AAB) have been investigated using density functional theory with a basis set that includes polarization functions on all the atoms. These aminoazo dyes are of interest because their carcinogenic activities depend dramatically on the position (n) of the methoxy group, e.g. 3-OMe-AAB is a potent hepatocarcinogen in the rat, whereas 2-OMe-AAB is a noncarcinogen. Although the various isomers of OMe-AAB require metabolic activation via N-hydroxylation prior to reaction with cellular macromolecules, we have shown that there are structural and electronic features present in these isomers that correlate with their carcinogenic behavior.

3-Methoxy-4-aminoazobenzene (3-OMe-AAB) is a potent hepatocarcinogen in the rat [1]. This aminoazo dye requires metabolic activation to N-hydroxy-3-methoxy-4-aminoazobenzene (N-OH-3-OMe-AAB) prior to reaction with cellular macromolecules [2]. This conclusion is in accord with the observation that 3-OMe-AAB is mutagenic on the Ames' *Salmonella* system only after treatment with S-9, the 9,000 g supernatant fraction of liver homogenate, whereas N-OH-3-OMe-AAB is strongly mutagenic without S-9 treatment [3,4]. Interestingly, changing the position of the methoxy group on the phenyl rings dramatically influences the carcinogenic behavior of the resulting compound [5]. For example, 2-OMe-AAB is noncarcinogenic in rats whereas 4′-OMe-AAB is carcinogenic, but to a lesser degree than 3-OMe-AAB. This carcinogenic potency of 2- and 4′-OMe-AAB correlates well with their mutagenic activity in the Ames' *Salmonella* test, where neither 2-OMe-AAB nor its N-hydroxy derivative, N-OH-2-OMe-AAB are mutagenic even after treatment with S-9; 4′-OMe-AAB is very slightly mutagenic on *Salmonella* (TA98) and N-OH-4′-OMe-AAB is definitely mutagenic without S-9 treatment [6]. Unfortunately, the carcinogen/mutagenic activities of the remaining monomethoxy derivatives of 4-aminoazobenzene or their N-hydroxy analogs have not been reported [7]. [1] For comparison, we note that the parent compound, 4-aminoazobenzene is only weakly carcinogenic in rats, nonmutagenic on *Salmonella* (TA98) with or without S-9 treatment and mutagenic on *Salmonella* (TA100) only in the presence of S-9 [6].

Although it is not entirely clear why there is such a radical difference in the carcinogenic behavior of 2- and 3-OMe-AAB, Kojima *et al.* [1] have determined that N-OH-3-OMe-AAB has a significantly greater effect than N-OH-2-OMe-AAB on DNA synthesis *in vivo*. This suggests that the observed differences in the carcinogenic activity of 2-OMe-AAB and 3-OMe-AAB may be linked to the differences in the inhibitory effects of their N-hydroxy derivatives on DNA replication. Hashimoto et al. [8] have established that the cytochrome P-450

---

[1] It is known that N,N-dimethyl-3′-OMe-AAB is carcinogenic

enzymes efficiently catalyze the mutagenic activation of 3-OMe-AAB and, in contrast to other carcinogenic aromatic amines, the activation is mediated by phenobarbital-P-450 rather than by 3-methyl-cholanthrene-P-450.

Despite significant interest in the carcinogenic behavior of the various positional isomers of OMe-AAB, relatively little is known about their structural or electronic properties. No experimental results from X-ray or electron diffraction studies have been reported for any of the OMe-AAB isomers [9]. Furthermore, no high-level computational results that compare the various OMe-AAB isomers using either molecular orbital or density functional theory calculations are currently available in the literature. It is important to note that substitution at the 2- and 6-positions or at the 3- and 5-positions in 4-aminoazobenzene are not equivalent, see Figure 1.

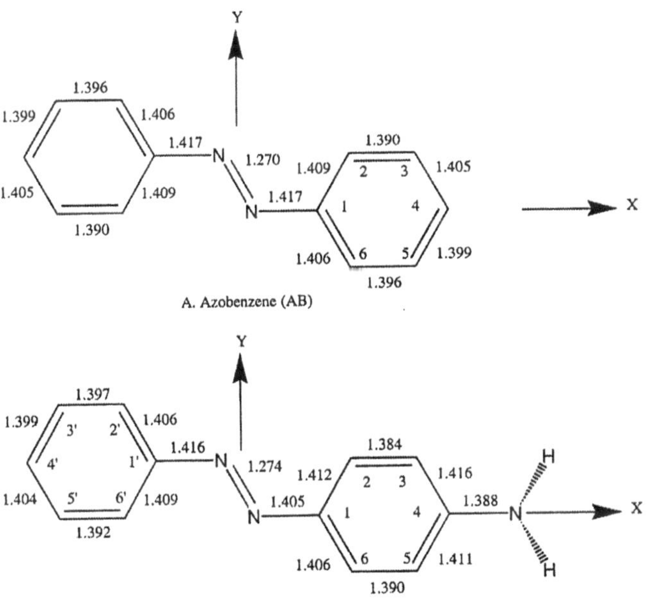

Figure 1. The structures, coordinate system and numbering conventions for A. Azobenzene (AB) and B. 4-Aminoazobenzene (AAB). The bond lengths (Å) shown were calculated at the BP/DN**//BP/DN** computational level

However, it is not evident that distinctions of this type were considered in the carcinogenic/mutagenic studies involving 2- and 3-OMe-AAB [1-5]. Thus, it is probably more appropriate to describe these studies as involving methoxy substitution at the meta and ortho positions respectively. The purpose of this chapter is to describe the results of an extensive computational study using density functional theory (DFT) to establish the conformational preferences and relative energies of the positional isomers of OMe-AAB. Our goal is to identify any electronic and/or structural features that may be present among these positional isomers that can be correlated with their diverse carcinogenic behaviors and lead to a better understanding of the underlying molecular mechanism(s) involved.

2 Computational Methods

Density functional calculations were performed at the BP/DN** computational level with SPARTAN v5.0 on Silicon Graphics computers [10]. This level uses the non-local Becke-Perdew (BP) 86 functional and employs the numerically defined DN** basis set which includes polarization functions on all the atoms [12].[2] Complete optimizations for a variety of conformers of each OMe-AAB derivative were carried out; no symmetry constraints were employed in order to minimize the likelihood of optimizing to a transition state. In a few cases frequency analyses were performed to ensure that the optimized structures were local minima on the potential energy surfaces (PESs). The graphics utilities of SPARTAN were used to examine the electron densities, electrostatic potentials and various Kohn-Sham orbitals for each conformer. Mulliken and electrostatic charges were also calculated.

2.1 Molecular Modeling Results and Discussion

Since no experimental structural data are available even for the parent compounds azobenzene (AB) and 4-aminoazobenzene (AAB), we first optimized these molecules at the BP/DN** computational level. The initial structures of both AB and AAB were taken as nearly *trans* about the azo linkage [13][3] and, in the case of AAB, the amino group was taken as pyramidal with both hydrogen atoms on the same side of the ring [14].[4] The optimized structure of AB is found to be planar

---

[2] Only a few experimental structures of azo dyes have been reported: O-Aminoazotoluene (X-ray) Kurosaki S., Kashino S., Haisa M. (1976), *Acta Cryst.* **B32**, pp. 3160; Disperse Red 167 (X-ray) Freeman H.S., Posey J.C. Jr., Singh P. (1992) *Dyes and Pigm.*, **20**, pp. 279; C.I. Disperse Yellow 86 (X-ray), Lye J., Hink D and Freeman H.S. (1997), *Comp. Chemistry applied to synthetic dyes.*

[3] A second conformer of AB, with the phenyl rings twisted some 80°, was found to be 2.3 kcal/mol higher in energy at the BP/DN** computational level.

[4] We also optimized a conformer of AAB in which the hydrogen atoms bonded to the amine nitrogen atom are on opposite sides of the ring but otherwise the structure is planar. This conformer is 8.0 kcal/mol higher in energy than the form shown in Figure 1. The

and a frequency analysis confirms that this is a local minimum on the PES. As can be seen in Figure 1, the azo-linkage in AB significantly distorts the carbon-carbon bond lengths in the phenyl rings compared to their values in benzene, where the carbon-carbon bond distances are 1.403Å at this computational level. The length of the N=N bond in AB, 1.270Å, suggests considerable electron delocalization; the N=N bond lengths in CH_3-N=N-CH_3 and CH_3-N=N-C_6H_5 are shorter, 1.250Å and 1.258Å respectively. In the optimized structure of AAB, the two phenyl rings are practically planar and nearly coplanar with each other, see Figure 1. A frequency analysis confirms that this is a local minimum on the PES. To a large extent, the calculated structural parameters of AAB show that the amine group at the 4-position reinforces the geometrical changes already induced by the azo linkage. This is a consequence of electron delocalization using the lone pair of electrons from the amine nitrogen atom to give the C_4-N bond partial double bond character, which results in predictable adjustments of the bond lengths in the remainder of the molecule. For comparison, we note that the lengths of the C-N bonds in CH_3-NH_2 and C_6H_5-NH_2 are 1.478Å and 1.408Å respectively, considerably longer than that found in AAB, 1.388Å. Nevertheless, the structure at the amine nitrogen in AAB remains pyramidal - the sum of the three bond angles is 346.7° at this computational level, compared to 318.4° and 325.6° for NH_3 and NH_2-CH_3 respectively.

It is of interest to compare a few of the Kohn-Sham molecular orbitals of AB and AAB. The highest occupied molecular orbital (HOMO) in AB is a lone-pair orbital localized primarily in the vicinity of the azo linkage, which is14.4 kcal/mol above the next highest occupied orbital (HOMO{-1}), a delocalized *pi*-bonding orbital. The lowest unoccupied molecular orbital (LUMO) in AB is a *pi*-antibonding orbital, some 46.7 kcal/mol above the HOMO. In AAB the HOMO also involves the azo lone-pair electrons, see Figure 2. It is nearly identical in shape to the HOMO found in AB, but it is 7.4 kcal/mol higher in energy. The HOMO{-1} in AAB involves the lone pair of electrons on the amine nitrogen atom, see Figure 2, but otherwise it is similar in shape to the HOMO{-1} (*pi*-bonding orbital) in AB. However, this HOMO{-1} is 18.6 kcal/mol above its counterpart in AB, reducing the energy gap between the two highest occupied orbitals in AAB to only 3.2 kcal/mol. The LUMO in AAB is similar in shape to the *pi*-antibonding LUMO in AB except that it includes a contribution from the amine nitrogen atom, see Figure 2. The energy gap between the HOMO and LUMO is 47.5 kcal/mol.

Two general types of conformers were considered for each of the positional isomers of monomethoxy AAB: one in which the O-Me bond is essentially in the nominal plane of the phenyl ring to which it is bonded and the other in which the

---

optimized structure of a completely planar form of AAB is a transition state that is 0.15 kcal/mol higher in energy than the lowest energy form shown in Figure 1.

O-Me bond is nearly perpendicular to this ring. For all nine isomers, the conformers in which the O-Me bond lies essentially in the ring plane are found to be *lower* in energy at the BP/DN**//BP/DN** computational level; the energies for these conformers are listed in Table 1 along with those for AB and AAB. Selected geometrical parameters and properties of the various OMe-AAB isomers calculated at the BP/DN** level are listed in Tables 2 and 3.

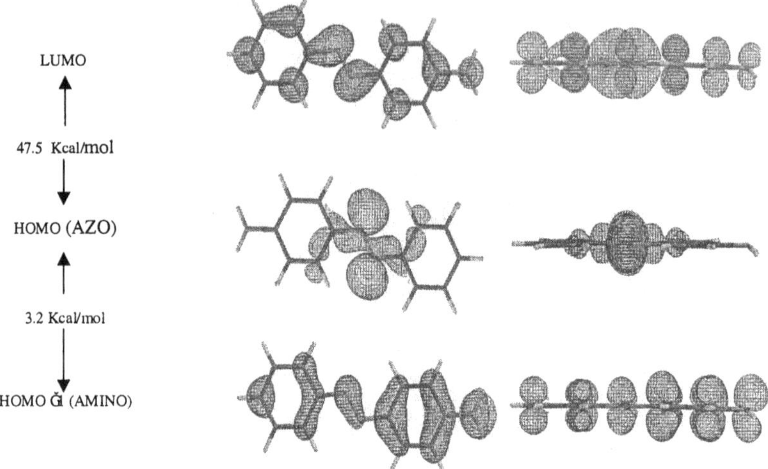

Figure 2 The HOMO{-1}, HOMO, and LUMO of 4-aminoazobenzene (AAB) calculated at the BP/DN**//BP/DN** computational level.

Since several orientations of the methoxy methyl group are possible, their positions for the lowest energy conformers at the BP/DN** level are shown in Figure 3. It should be noted that energy differences between some of the conformers of these monomethoxy isomers can be quite small. For example, rotating 180° about the C_ϕ-O bond in 4′-OMe-AAB and re-optimizing yields a structure only 0.3 kcal/mol higher in energy, whereas a conformer with the methyl group nearly perpendicular to the ring is 3.8 kcal/mol higher in energy.

2.1.1 3-and 5-OMe-AAB

As can be seen from Table 1, 3-OMe-AAB has the lowest total molecular energy among all the positional isomers at the BP/DN**//BP/DN** computational level. However, the other ortho derivative, 5-OMe-AAB, is less than 1 kcal/mol higher in energy. As expected, the presence of a methoxy group ortho to the amine group perturbs the pattern of carbon-carbon bond lengths in the phenyl rings compared to those in AAB. The most prominent changes occur at the point of attachment, e.g. in the case of 3-OMe-AAB, the length of the shorter bond (C_2-C_3) decreases while the length of the longer bond (C_3-C_4) increases, see Table 2. As might be expected, the lengths of the bonds in the unsubstituted phenyl ring are not significantly altered by the presence of the methoxy group. In both 3- and 5-OMe-AAB the length of C_4-N bond is shorter than that found in AAB and the amine group is less pyramidal, see Table 2. This suggests a further delocalization of the lone pair electron density on the amine nitrogen atom; the calculated Mulliken and electrostatic charges on this nitrogen are less negative than those found in AAB. The C_ϕ-O bond lengths in the 3- and 5-isomers, 1.378Å and 1.380Å respectively, are *longer* than those calculated for the other positional isomers, see Table 2. In order to examine the effect of the amine group on the length of the C_ϕ-O bond, we replaced the amine group in 3-OMe-AAB with a hydrogen atom and reoptimized the structure. In the resulting 3-OMe-AB compound (as well as in MeO-C_6H_5) the C_ϕ-O bond length, 1.373Å, is slightly shorter than that found in 3-OMe-AAB. Thus, the amine group at the 4-position tends to impede the delocalization of lone-pair density on the methoxy oxygen atom in 3-OMe-AAB.

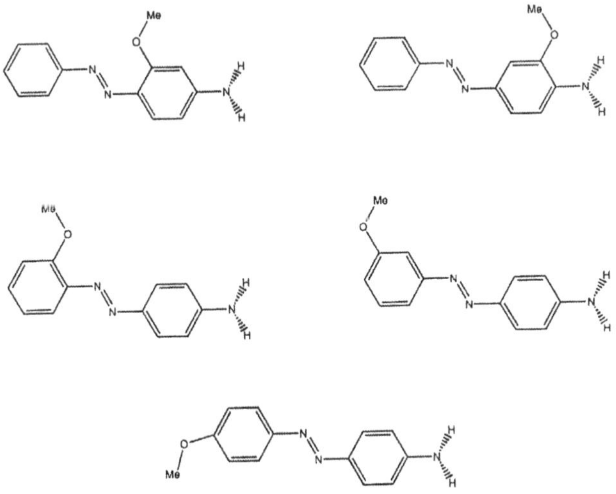

Figure 3. The orientation of the methyl group for the positional isomers of monomethoxy-4-aminobenzene calculated at the BP/DN**//BP/DN** computational level. (The orientation of the methyl group in 6-OMe-AAB is analogous to that for 2-OMe-AAB, etc.)

The presence of a methoxy group ortho to the amine group in AAB has an interesting effect on the two highest occupied Kohn-Sham molecular orbitals. The orbital localized at the azo linkage in 3(5)-OMe-AAB is similar in shape and only 0.9(1.5)kcal/mol higher in energy than the corresponding orbital in AAB; the in-plane lone pair on the methoxy oxygen is represented in this orbital, but only to a small extent. The orbital involving the amine lone pair of electrons, which includes a significant contribution from the out-of-plane oxygen lone pair, is 5.3(5.7) kcal/mol higher in energy than the corresponding orbital in AAB. These relatively large increases in energy make this orbital the HOMO in both 3- and 5-

OMe-AAB. It is important to note that methoxy substitution at each of the positions on the phenyl rings increases the energy of this orbital, but the increases for the 3- and 5-isomers are more than triple the smallest increase we observed, 1.7 kcal/mol for 5′-OMe-AAB. The relatively large increase in the energy of this orbital appears to be a result of electron overcrowding involving the proximate lone pairs on the methoxy oxygen and amine nitrogen atoms. The shape of the LUMO in 3(5)-OMe-AAB is quite similar to the LUMO in AAB, having relatively little contribution from the out-of-plane lone-pair orbital on the methoxy oxygen atom. Furthermore, the energy of these LUMO in both the 3- and 5-isomers is less than 2 kcal/mol above the LUMO in AAB. The energies separating the HOMO and LUMO in 3- and 5-OMe-AAB, 47.2 and 46.5 kcal/mol, are just slightly lower than that found in AAB.

Table 1 Total Molecular Energies (a.u.) of n-Methoxy-4-Aminoazobenzene calculated at the BP/DN**//BP/DN** Computational Level.

n	Total Molecular Energies (BP/DN **//BP/DN**)	Relative Energies
2	-742.931442	+6.7
3	-742.942450	0.0
5	-742.941042	+0.9
6	-742.936054	+4.0
2′	-742.935475	+4.4
3′	-742.940108	+1.5
4′	-742.940449	+1.3
5′	-742.941441	+0.6
6′	-742.930842	+7.3
AAB	-628.365269	-
AB	-572.974339	-

2.1.2 2- and 6-OMe-AAB

The positional isomers 2- and 6-OMe-AAB are 6.7 and 4.0 kcal/mol higher in energy than the lowest energy isomer, 3-OMe-AAB, at the BP/DN**//BP/DN** computational level, see Table 1. The length of the C_4-N bond in both of these isomers is slightly longer than that found in AAB and the amine group is more pyramidal, see Table 2. The calculated Mulliken and electrostatic charges on the amine nitrogen atom in the 2- and 6-isomers are nearly the same as those in AAB. However, the C_ϕ-O bond lengths in 2(6)-OMe-AAB, 1.356Å(1.362Å), are some

0.02Å shorter than the corresponding bond lengths in 3(5)-OMe-AAB. This indicates further delocalization involving the oxygen out-of-plane lone pair, which gives the C_ϕ-O bond additional double bond character. The Mulliken charge on the oxygen atom in 2-OMe-AAB is not as negative as that found in 3-OMe-AAB. To examine the effect of the amine group on the length of the C_ϕ-O bond in 2-OMe-AAB, we optimized the structure of the 2-OMe-AB. The length of the C_ϕ-O bond increases, but only by about 0.002Å; this change, however, is in a direction opposite to what we observed in going from 3-OMe-AAB to 3-OMe-AB. The shorter C_ϕ-O bond in 2(6)-OMe-AAB results in an elongation of both carbon-carbon bonds in the ring at the point of methoxy attachment when compared to that in AAB, see Table 2. Again, there are no significant changes in the carbon-carbon bond lengths in the unsubstituted phenyl ring compared to those in AAB.

The presence of a methoxy group meta to the amine group in AAB alters the two highest occupied Kohn-Sham orbitals differently than when the replacement occurs at an ortho position. In particular, the azo lone-pair orbital in 2(6)-OMe-AAB is 10.0(5.3) kcal/mol higher in energy than the corresponding orbital in AAB, but only 0.9(1.5) kcal/mol higher for substitution at the 3(5)-position. This orbital involves contributions from the azo nitrogen lone pairs and from the in-plane methoxy oxygen lone-pair; its relatively large increase in energy is clearly the result of adverse lone-pair interactions in the region. The particular geometrical arrangement of atoms in the vicinity of the *trans* azo linkage causes the electron overcrowding in this region to be more severe for methoxy substitution at the 2-position than at the 6-position. This leads to a greater increase in the energy of the azo lone pair orbital in 2-OMe-AAB and results in the largest energy gap between the two highest occupied orbitals we observed in this study, 9.0 kcal/mol. The orbital involved with the amine nitrogen lone pair in 2(6)-OMe-AAB is similar in shape to the corresponding orbital in AAB, although it includes a contribution from the out-of-plane oxygen lone pair. Its energy is raised to a slightly lesser extent than it is when the methoxy group is at an ortho position. Thus, for 2(6)-OMe-AAB and AAB the orbital involving the azo lone pair is higher in energy than the orbital involving the amine nitrogen lone pair, whereas for 3(5)-OMe-AAB the order of these two orbitals is reversed. It is also interesting to note that the energy gap between the two highest occupied molecular orbitals in 2(6)-OMe-AAB, 9.0(3.6) kcal/mol, is greater than that in AAB, 3.2 kcal/mol, and in 3(5)-OMe-AAB, 1.3(1.1) kcal/mol. The LUMO in both the 2- and 6-isomers is similar in shape to that in AAB, but involves a significant contribution from the out-of-plane lone-pair orbital on the methoxy oxygen atom. The LUMO energies of 2- and 6-OMe-AAB are higher than those of 3- and 5-OMe-AAB, whereas the separation in energy between the HOMO and LUMO is smaller, 44.1 and 45.6 kcal/mol respectively.

2.1.3 3′- and 5′-OMe-AAB

The positional isomers 3′(5′)-OMe-AAB are only 1.5 (0.6) kcal/mol higher in energy than 3-OMe-AAB, see Table 1. Interestingly, methoxy substitution at the 3′- or 5′-position has very little effect on the carbon-carbon bond lengths in either of the phenyl rings when compared to those in AAB, see Table 2. The C_4-N bond length in 3′-OMe-AAB is slightly shorter than that found in AAB while that of 5′-OMe-AAB is nearly the same as in AAB. The small differences in the geometrical parameters and the lack of any severe lone-pair interactions in 3′(5′)-OMe-AAB are consistent with the observation that the two highest occupied Kohn-Sham orbitals of AAB are closer in energy to those of 3′(5′)-OMe-AAB than to those of the other positional isomers. The azo and amine lone-pair orbitals are only 1.0(0.3) and 2.4(1.7) kcal/mol higher in energy than the corresponding orbitals in AAB, leading to an a small energy gap of 1.8(1.7) kcal/mol. The LUMOs are again similar in shape to that found in AAB, with relatively little contribution from the out-of-plane oxygen lone-pair orbital; the HOMO-LUMO energy gaps, 47.6 and 48.0 kcal/mol, are slightly greater than that found in AAB.

2.1.4 2′- and 6′-OMe-AAB

The positional isomers 2′- and 6′-OMe-AAB are found to be 4.4 and 7.3 kcal/mol higher in energy than 3-OMe-AAB, see Table 1. As can be seen in Table 2, methoxy substitution at the 6′-position of AAB leads to greater changes in several of the geometrical parameters than those found with the other monomethoxy derivatives. The relatively short $C_{6'}$-O bond length, 1.359Å, in 6′-OMe-AAB indicates significant electron donation from the out-of-plane lone pair on the methoxy oxygen. This short $C_{6'}$-O bond is compensated for by an elongation of both carbon-carbon bonds in the ring involving the $C_{6'}$ atom, a shortening of the $C_{1'}$-N bond and a slight elongation of the N=N bond. Analogous changes in the bond lengths occur for substitution at the 2-position, but the presence of the amine group buffers the magnitude of these changes somewhat, particularly at the azo linkage, see Table 2. The energies of the highest two Kohn-Sham molecular orbitals in AAB are both significantly increased by substitution at the 6′-position.

The lone-pair orbital localized at the azo linkage also includes a contribution from the in-plane oxygen lone pairs and remains higher in energy than the orbital involving the amine nitrogen lone pair; the energy separation, 5.8 kcal/mol, is the second highest we observed for any of the monomethoxy derivatives. The structure of the LUMO is similar to that in AAB, but contains a contribution from the out-of-plane oxygen lone-pair, similar to that observed for the 2- and 6-isomers. The energy separations between the HOMO and LUMO are 45.1 and 44.6 kcal/mol respectively.

2.1.5 4′-OMe-AAB

The positional isomer 4′-OMe-AAB is only 1.3 kcal/mol higher in energy than 3-OMe-AAB, see Table 1. As can be seen in Table 2, the $C_{4'}$-O bond length,

1.370Å, is intermediate between that found for the 3-, 3'-, 5- and 5'-isomers and that found for the 2-, 2'-, 6- and 6'-isomers.

Table 2 Structural Parameters (bond lengths (Å), bond angles (°)) of AB, AAB and n-OMe-AAB

N	C_1-C_2	C_2-C_3	C_3-C_4	C_4-C_5	C_5-C_6	C_6-C_1	C_4-N	C_1-N	N = N
2	1.436	1.398	1.409	1.409	1.386	1.410	1.391	1.396	1.278
3	1.416	1.382	1.428	1.404	1.393	1.403	1.380	1.402	1.278
5	1.407	1.388	1.409	1.421	1.388	1.412	1.384	1.404	1.275
6	1.409	1.384	1.415	1.408	1.400	1.427	1.390	1.396	1.277
2′	1.411	1.384	1.415	1.410	1.392	1.407	1.392	1.406	1.277
3′	1.411	1.383	1.416	1.411	1.390	1.408	1.385	1.405	1.274
4′	1.412	1.385	1.415	1.410	1.390	1.407	1.390	1.404	1.276
5′	1.411	1.384	1.416	1.410	1.389	1.408	1.388	1.401	1.274
6′	1.412	1.383	1.414	1.410	1.392	1.406	1.392	1.412	1.277
AAB	1.412	1.384	1.416	1.411	1.390	1.406	1.388	1.405	1.274
AB	1.409	1.390	1.405	1.399	1.396	1.406	-	1.417	1.270

n	$C_{1'}$-N	$C_{1'}$-$C_{2'}$	$C_{2'}$-$C_{3'}$	$C_{3'}$-$C_{4'}$	$C_{4'}$-$C_{5'}$	$C_{5'}$-$C_{6'}$	$C_{6'}$-$C_{1'}$	$C_{\emptyset}$-O	O-C	Σ angle [a]
2	1.414	1.406	1.395	1.399	1.404	1.392	1.408	1.356	1.430	344.5
3	1.413	1.405	1.396	1.399	1.404	1.392	1.409	1.378	1.433	348.9
5	1.417	1.406	1.396	1.398	1.403	1.392	1.410	1.380	1.431	346.8
6	1.416	1.406	1.397	1.399	1.404	1.392	1.410	1.362	1.430	346.0
2′	1.404	1.425	1.404	1.398	1.401	1.392	1.406	1.364	1.431	344.6
3′	1.416	1.408	1.398	1.406	1.398	1.395	1.405	1.374	1.433	349.2
4′	1.411	1.408	1.389	1.406	1.408	1.391	1.406	1.370	1.432	345.7
5′	1.415	1.402	1.399	1.393	1.411	1.394	1.411	1.373	1.432	347.2
6′	1.401	1.410	1.390	1.397	1.398	1.404	1.432	1.359	1.433	345.2
AAB	1.416	1.406	1.397	1.399	1.404	1.392	1.409	-	-	346.7
AB	1.417	1.406	1.396	1.399	1.405	1.390	1.404	-	-	-

[a] Sum of three bond angles at the amine nitrogen atom.

The pattern of carbon-carbon bond lengths in the phenyl ring to which the methoxy group is attached is generally enhanced above that already found in AAB. The structures of the two highest occupied Kohn-Sham molecular orbitals are radically different from those observed for the other positional isomers. They are nearly degenerate (only separated by 0.5 kcal/mol) and appear as a mixture of the orbitals involving the azo and the amine lone pairs that are found for the other positional isomers. For comparison, we optimized several other AAB derivatives with substitution at the 4'-position. Similar combination orbitals are obtained for the HOMO and HOMO{-1} of 4′-OH-AAB, but for 4′-F-AAB the HOMO is clearly an azo lone-pair type orbital, whereas for 4′-SMe-AAB the HOMO is an amine lone-pair type orbital. The structure of the LUMO in 4'-OMe-AAB is similar to that observed for AAB, but with a contribution from the out-of-plane lone-pair orbital on the oxygen atom; the energy gap between the HOMO and LUMO is 48.6 kcal/mol.

Table 3 Selected properties of AB, AAB, and n-OMe-AAB calculated at the BP/DN**//BP/DN** Computational Level.

n	HOMO{-1} (a.u.)	HOMO (a.u.)	LUMO (a.u.)	Log P[d]	Electrostatic Charge On Amine Nitrogen	Dipole Moment (D)
2	-0.186080[a]	-0.171712[b]	-0.101479	2.25	-0.72	4.85
3	-0.186278[b]	-0.184263[a]	-0.109082	2.13	-0.67	3.52
5	-0.185290[b]	-0.183553[a]	-0.109506	2.19	-0.67	3.54
6	-0.185048[a]	-0.179274[b]	-0.106652	2.26	-0.72	5.08
2′	-0.183749[a]	-0.179501[b]	-0.107565	2.29	-0.72	2.77
3′	-0.188844[a]	-0.186038[b]	-0.110120	2.24	-0.73	4.54
4′	-0.183084[c]	-0.182205[c]	-0.104728	2.34	-0.72	2.55
5′	-0.189970[a]	-0.187243[b]	-0.110805	2.53	-0.71	4.50
6′	-0.183052[a]	-0.173803[b]	-0.102735	2.30	-0.72	2.04
AAB	-0.192206[a]	-0.187648[b]	-0.111910	2.47	-0.72	3.61
AB	-0.222337[a]	-0.199380[b]	-0.126611	3.30	-	0.07

a. Orbital involves amine lone pair.
b. Orbital involves azo lone pairs.
c. Orbital is mixed, see text.
d. Log P is the logarithm of the octanol-water partition coefficient calculated using the Dixon-Hehre algorithm in Spartan 5.0 [10]. This involves explicit evaluation of $AM1_{oct}$ and $AM1_{aq}$ solvation models. The Ghose-Crippen approach gives Log P = 3.54 for all the OMe-AAB isomers (*J. Comp. Chem.* 9, 80 (1988).

2.2 Remarks

The methoxy azo dyes 2-OMe-AAB, 4′-OMe-AAB and 3-OMe-AAB are noncarcinogenic, moderately carcinogenic and strongly carcinogenic respectively.[5] The studies that established these results, however, have not made a clear distinction between methoxy substitution at the 2- and 6-position or at the 3- and 5-position. Ames' Salmonella mutagenicity tests suggest that none of these molecules are mutagenic per se, but require activation to their N-hydroxy derivatives prior to reaction with cellular macromolecules. Nevertheless, there appear to be some differences in the structures and electronic properties of the monomethoxy-AAB compounds themselves that may provide a basis for understanding their diverse carcinogenic behavior.

Many of the structural features in the monomethoxy AAB derivatives are determined to a large extent by electron delocalization at the azo linkage, which establishes a pattern of carbon-carbon bond lengths in the phenyl rings of AB; this pattern is enhanced by electron delocalization at the amine nitrogen atom in AAB. The presence of a methoxy group, with its two lone pairs of electrons, provides yet another site where delocalization is an issue, but it also introduces the possibility of lone-pair interactions involving the azo and amine nitrogen lone pairs.

Comparing the structures of 3(5)-OMe-AAB with those of 2(6)-OMe-AAB suggests that there is competition to delocalize lone-pair electron density at the amine nitrogen and methoxy oxygen atoms. For the 3(5)-isomers, where the methoxy oxygen is in close proximity to the amine nitrogen, it is energetically favorable to delocalize at the (less-electronegative) amine nitrogen atom by further increasing the double bond character of the C_4-N bond; for these isomers the C_ϕ-O bond is relatively long. For the 2(6)-isomers, where the methoxy oxygen is now in close proximity to the azo linkage, it becomes favorable to delocalize more at the methoxy oxygen atom by further increasing the double bond character of the C_ϕ-O bond; for these isomers the C_4-N bond is relatively long.

The Kohn-Sham HOMO and HOMO{-1} of AAB involve the azo and amine lone pairs respectively and these orbitals are relatively close in energy. For most of the monomethoxy AAB derivatives the two highest occupied orbitals are similar in shape to those found in AAB, but involve contributions from one of the two lone pairs on the methoxy oxygen atom. In these n-OMe-AAB compounds, the energies of the two highest occupied orbitals are sensitive to the position (n) of the methoxy group because there is the potential for its lone pairs to be forced into close proximity with those on the AAB backbone.

It is interesting to note that the HOMO of the strongest carcinogen, 3(5)-OMe-AAB, involves the amine lone pair, whereas the HOMO of the noncarcinogen, 2(6)-OMe-AAB, involves the azo lone pairs. In the case of AAB itself, which is weakly carcinogenic, the HOMO involves the azo lone pairs, but the separation in energy between the two highest orbitals is smaller than that for 6-OMe-AAB and much smaller than that for 2-OMe-AAB. The carcinogenic potency of 4′-OMe-AAB is in between that of 2(6)-OMe-AAB and 3(5)-OMe-AAB and its HOMO is a mixed orbital that includes a contribution from the amine nitrogen lone pair.

The results of our investigation suggest that the carcinogenic activity of an OMe-AAB isomer is increased as the energy of the orbital involving the amine nitrogen lone pair is raised relative to that of the orbital involving the azo nitrogen lone pairs.[5] This correlation can be further tested by noting that N-methyl-AAB and N,N-dimethyl-AAB compounds are usually more carcinogenic than the corresponding AAB compounds [6]. The HOMOs of both AAB and N-methyl-AAB are localized at the azo linkage. However, the separation in energy between the HOMO and HOMO{-1} in N-methyl-AAB is only about 50% of the corresponding separation in AAB. On the other hand, the HOMOs of both 3-OMe-AAB and N-methyl-3-OMe-AAB involve the amine lone pair, but for these compounds the energy gap between the two highest occupied orbitals is three times greater in the N-methyl compound. Furthermore, the HOMO and HOMO{-1} of N,N-dimethyl-AAB are of the mixed type we observed in 4′-OMe-AAB, where both orbitals involve the amine nitrogen lone pair. These results are consistent with an increase in the carcinogenic potency of a methoxy AAB derivative when the primary amine is monomethylated or dimethylated. It must be pointed out that a variety of effects can influence the carcinogenic activity of a particular compound [13-15]. For example, the HOMO of N,N-dimethyl-4′-OH-AAB involves the amine nitrogen lone pair and it is 2.7 kcal/mol higher in energy than the orbital involving the azo lone pairs. Based on our results for 2- and 3-OMe-AAB, this would suggest that N,N-dimethyl-4′-OH-AAB was a strong hepatocarcinogen in the rat, but this is not the case [16]. It is likely that the hydroxy group provides a site for the metabolic breakdown of this dye before it can act as a carcinogen [16]. In fact, studies have shown that N,N-dimethyl-4′-OH-AAB is formed from N,N-dimethyl-AAB during its metabolism by rat homogenates [17], and that demethylated hydroxyazo derivatives are present in the urine of rats fed the dye [18]. Additional calculations and further experimental carcinogenic/mutagenic studies on AAB derivatives will be required to establish the extent to which knowing the relative energies of the orbitals involving the azo and amine lone pairs in these compounds can be used as a predictive tool of the carcinogenic behavior of azo dyes. These studies are currently in progress.

3 Neural Network Approach

In the last several years there has been a large and energetic upswing in research efforts aimed at synthesizing fuzzy logic with computational neural networks in the emerging field of soft computing in AI. The enormous success of commercial applications (primarily by Japanese companies), which are dependent to a large

---

[5] The energies of the LUMOs do not seem to correlate well with the carcinogenic potency of the AAB compounds, e.g. the LUMO of the noncarcinogen 2-OMe-AAB is 4.4 kcal/mol above the LUMO of the strong carcinogen 3-OMe-AAB but 6.5 kcal/mol above the LUMO of the weak carcinogen AAB.

extent on soft computing technologies, has led to a surge of interest in these techniques for possible applications throughout the US textile industry.

The marriage of fuzzy logic with computational neural networks has a sound technical basis, because these two approaches generally attack the design of "intelligent" systems from quite different angles. Neural networks are essentially low level, computational algorithms that offer good performance in dealing with large quantities of data often required in pattern recognition and control.

Fuzzy logic, introduced in 1965 by Zadeh [19] is a means for representing, manipulating and utilizing data and information that possess non-statistical uncertainty. Thus, fuzzy methods often deal with issues such as reasoning on a higher (i.e., on a semantic or linguistic) level than do neural networks.

Consequently, the two technologies often complement each other: neural networks supply the brute force necessary to accommodate and interpret large amounts of data and fuzzy logic provides a structural framework that utilizes and exploits these low level results.

This research is concerned with the integration of fuzzy logic and computational neural networks. Therefore, an algorithm for the creation and manipulation of fuzzy membership functions, which have previously been learned by a neural network from data set under consideration, is designed and implemented. In the opposite direction we are able to use fuzzy tree architecture to construct neural networks and take advantage of the learning capability of neural networks to manipulate those membership functions for classification and recognition processes.

In this research, membership functions are used to calculate fuzzy entropies for measuring uncertainty and information. That is, the amount of uncertainty regarding some situation represents the total amount of potential information in this situation. The reduction of uncertainty by a certain amount (due to new evidence) indicates the gain of an equal amount of information.

3.1 Fuzzy Entropy Measures

In general, a fuzzy entropy measure is a function $f: P(X) \rightarrow R$, where P(X) denotes the set of all fuzzy subsets of X. That is, the function f assigns a value f(A) to each fuzzy subset A of X that characterizes the degree of fuzziness of A. Thus, f is a set-to-point map, or in other words, a fuzzy set defined on fuzzy sets [20].

DeLuca and Termini [21] first proposed the axioms of non-probabilistic entropy. Their axioms are intuitive and have been widely accepted in the fuzzy literature. We adopt them here. In order to qualify as a meaningful measure of fuzziness, f must satisfy the following axiomatic requirements:

Axiom 1. $f(A) = 0$ if and only if A is a crisp (non-fuzzy) set.

Axiom 2. f(A) assumes the maximum if and only if A is maximally fuzzy.

Axiom 3. If A is less fuzzy than B, then f(A) ≤f(B).

Axiom 4. f(A) = f(A^C).

Only the first axiom is unique; axioms two and three depend on the meaning given to the concept of the degree of fuzziness. For example, assume that the "less fuzzy" relation is defined, after DeLuca and Termini [21], as follows:

$$\mu_A(x) \leq \mu_B(x) \text{ for } \mu_B(x) = ½1/2 \tag{1}$$

$$\mu_A(x) \geq \mu_B(x) \text{ for } \mu_B(x) = ½1/2, \tag{2}$$

and the term maximally fuzzy is defined by the membership grade 0.5 for all x ∈X.

Motivated by the classical Shannon entropy function DeLuca and Termini proposed the following fuzzy entropy function [21]:

$$f(A) = - \Sigma(\mu_A (x)\log_2\mu_A (x) + (1 - (\mu_A (x))\log_2(1 - \mu_A (x))). \tag{3}$$

Its normalized version is given by f(A)/ |X|, where |X| denotes the cardinality of the universal set X. Similarly, taking into account the distance from set A to its complement A^C another measure of fuzziness, referred to as an index of fuzziness [22], can be introduced. If the "less fuzzy" relation of Axiom 3 is defined by:

$$\mu_C(x) = 0 \text{ if } \mu_A(x) = ½, \tag{4}$$

$$\mu_C(x) = 1 \text{ if } \mu_A(x) > ½, \tag{5}$$

where C is the crisp set nearest to the fuzzy set A, then the measure of fuzziness is expressed by the function [22]:

$$f(A) = \Sigma(\mu_A (x) - \mu_C(x)) \tag{6}$$

when the Hamming distance is used, and by the function [22]:

$$f(A) = (\Sigma(\mu_A (x) - \mu_C(x))^2)^{1/2} \tag{7}$$

when the Euclidean distance is employed.

It is clear that other metric distances may be used as well [23]. For example, the Minkowski class of distances yields a class of fuzzy measures:

$$f_w(A) = (\Sigma(\mu_A (x) - \mu_C(x))^w)^{1/w} \tag{8}$$

where w $\in[1, \infty)$.

However, both DeLuca and Termini measure, and Kaufmann measure are only special cases of measures suggested by Knopfmacher [24] and Loo [25], expressed in the form [23]:

$$f(A) = h(\Sigma g_x(\mu_A(x))), \tag{9}$$

where $g_x (\mu_A(x))$ are functions

$$g_x : [0, 1] \rightarrow R^+ , \tag{10}$$

which are all monotonically increasing in [0, 0.5], monotonically decreasing in [0.5, 1], and satisfy the requirements that $g_x(0) = g_x(1) = 0$, and that $g_x(0.5)$ is the unique maximum of g_x, and h is a monotonically increasing function. It has been shown that the degree of fuzziness of a fuzzy set can be expressed in terms of the lack of distinction between the set and its complement [26-28].

It has been also established that a general class of measures of fuzziness based on this lack of distinction is exactly the same as the class of measures of fuzziness expressed in terms of a metric distance based on some form of aggregating the individual differences [23]:

$$f_C(A) = |X| - \Sigma(\mu_A (x) - c(\mu_A(x))). \tag{11}$$

To obtain the normalized version of a fuzzy entropy the above expression is divided by the cardinality of a fuzzy set. The previous definitions can also be extended to infinite sets [23].

Another fuzzy entropy measure was proposed and investigated by Kosko [20,29]. He established that

$$f(A) = (\Sigma count\ (A \wedge A^C))/\ (\Sigma count\ (A \vee A^C)), \tag{12}$$

where Σcount (sigma-count) is a fuzzy cardinality [30,31].

Kosko [29] claims that his entropy measure, and corresponding fuzzy entropy theorem does not hold when we substitute Zadeh's operations [19] with any other generalized fuzzy operations. If any of the generalized Dombi's operations [32] are used, the resulting measure is an entropy measure, it is maximized, however it does not equal unity at the midpoints [33].

The generalized Dombi's operations proved to do well in different applications, and were used by Sztandera [34] for detecting coronary artery disease, and were suggested for image analysis by Sztandera [35]. However, we still have to use Zadeh's complement [19], since the Kosko's theorem does not hold for any other class of fuzzy complements.

3.2 A New Concept of Fuzzy Entropy

In our experiments we used fuzzy entropy suggested by Kosko [29] and generalized fuzzy operations introduced by Dombi [32].

Generalized Dombi's operations form one of the several classes of functions, which possess appropriate axiomatic properties of fuzzy unions and intersections. The operations are defined below. From our experience the parameter $\lambda = 4$ gives the best results [33].

3.2.1 Dombi's Fuzzy Union

$$\mu_{A \vee B}(x) = \{1 + [(1/\mu_A(x) - 1)^{-\lambda} + [(1/\mu_B(x) - 1)^{-\lambda}]^{-1/\lambda}]\}^{-1} \tag{13}$$

where λ is a parameter by which different unions are distinguished, and $\lambda \in (0,\infty)$.

3.2.2 Dombi's Fuzzy Intersection

$$\mu_{A \wedge B}(x) = \{1 + [(1/\mu_A(x) - 1)^{\lambda} + [(1/\mu_B(x) - 1)^{\lambda}]^{1/\lambda}]\}^{-1} \tag{14}$$

where λ is a parameter by which different intersections are distinguished, and $\lambda \in (0,\infty)$.

It is interesting to examine the properties of these operations. By definition, generalized fuzzy union and intersection operations are commutative, associative, and monotonic. It can be shown that they neither satisfy the law of the excluded middle nor the law of contradiction. They are also not idempotent, nor distributive. However, they are continuous and satisfy the de Morgan's laws (when the standard Zadeh's complement is used) [32]. Zadeh's complement ($c(a) = 1 - a$) is by definition monotonic non-increasing. It is also continuous and involutive. Other properties and the proofs can be found in Dombi's [32] and Zadeh's [19] papers.

4 Feed-Forward Neural Network Architecture

The proposed algorithm generates feed forward network architecture for a given data set, and after having generated fuzzy entropies at each node of the network, it switches to fuzzy decision making on those entropies. The nodes and hidden layers are added until a learning task is accomplished. The algorithm operates on numerical data and equates a decision tree with a hidden layer of a neural network [33]. A learning strategy used in this approach is based on achieving the optimal goodness function. This process of optimization of the goodness function translates into adding new nodes to the network until the desired values are achieved. When this is the case then all training examples are regarded as correctly recognized. The incorporation of fuzzy entropies into the algorithm seems to result in a drastic reduction of the number of nodes in the network, and in decrease of the convergence time. Connections between the nodes have a "cost" function being equal to the weights of a neural network. The directional vector of a hyper-plane, which divides decision regions, is taken as the weight vector of a node.

The outline of the algorithm follows:

Step i) For a given problem with N samples, choose a random initial weight vector.

Step ii) Make use of learning rule

$\Delta w_{ij} = -\rho\, \partial f(F)/\partial w_{ij}$

where ρ is a learning rate, $f(F)$ is a fuzzy entropy function; and search for a hyper-plane that minimizes the fuzzy entropy function:

$\min f(F) = \Sigma N_r/N \text{ entropy}(L, r)$

where: L is a level of a decision tree, R is total number of nodes in a layer, r is number of nodes, f (F) is fuzzy entropy.

Step iii) If the minimized fuzzy entropy is not zero, but it is smaller than the previous value compute a new node in the current layer and repeat the previous step. Otherwise go to the next step.

Step iv) If there is more than one node in a layer compute a new layer with inputs from all previous nodes including the input data, then go to step ii). Otherwise terminate.

5 Azo Dye Database

We have conclusively demonstrated that density functional techniques can efficiently be used to investigate the structure and properties (charge distribution, band gap, log P, etc.) of a wide range of azo dyes. (Most prior calculations on dyes have used lower level semi-empirical methods). We employed the gradient-corrected density functional (Becke-Perdew) method incorporated into the Spartan 5.0 molecular modeling package [10] using the polarized numerical DN** basis set (BP/DN**//BP/DN** level), which provides an exceptionally good description of the bonding in most organic molecules. (This computational level can also be used with dyes that contain metals such as Cr, Co, Cu, etc.). The calculated structural and physicochemical properties of these dyes, augmented with experimental results (optical properties, toxicological activity, etc.) were incorporated into a database that was used to train the neural network.

Preliminary results from several trials suggest that, given a collection of dye molecules, each described by a set of structural features, a set of physical properties, and the strength of some activity under consideration, a neural network algorithm could be used to find patterns in terms of the structural features and properties that correspond to a desired level of activity.

To determine the effectiveness of the proposed algorithm, the performance was evaluated on a database of molecular properties involving 22 selected azo dyes (11 carcinogenic/mutagenic and 11 non-carcinogens). We used 80% of the database (18 molecules) for training purposes, and 20% (4 molecules) for testing. We repeated the process five times (20% Jackknife procedure). After several trial-and-error approaches with different input sets, we opted for three input parameters (log P, surface, and volume). Using those parameters, in conjunction with experimental toxicological data, the network was able to learn and differentiate between mutagenic/carcinogenic and non-mutagenic/non-carcinogenic dyes. We expect the neural network to predict the mutagenic/carcinogenic nature of other chemical structures.

We are currently looking into using so called topological indices (modified Wiener's index, modified Balaban's index, modified Schultz's index, etc.) [36, 37] that have been used successfully in the pharmaceutical industry in QSPR and QSAR studies. We plan to use one of these topological indices, or develop one ourselves if none of these are adequate, in conjunction with log P and selected electronic properties from our density functional calculations as descriptors in our soft computing approach.

6 Concluding Remarks

From the soft computing point of view, the proposed approach shows a way in which neural network technology can be used as a "tool" within the framework of a fuzzy set theory. Generating membership functions with the aid of a neural network has been shown to be an extremely powerful and promising technology. In this research, membership functions are used to calculate fuzzy entropies for measuring uncertainty and information. The proposed neural network is a building block towards combining the two soft computing paradigms. It allows for a self-generation of a neural network architecture suited for a particular problem.

The main features and advantages of the proposed approach are: 1) it is a general method of how to use numerical information, via neural networks, to provide good approximations to the membership functions; 2) it is a simple and straightforward quick-pass build-up procedure, where no time-consuming iterative training is required, resulting in much shorter design time than most neural networks; 3) there is a lot of freedom in choosing the membership functions and corresponding fuzzy entropies; this provides flexibility for designing systems satisfying different requirements; and 4) it performs successfully on data where neither a pure neural network nor a fuzzy system would work perfectly.

Molecular modeling has allowed us to investigate the properties of a large number of azobenzene derivatives in a short period of time. It is clear that there are correlations between our calculated properties and their toxicological behavior. We are also certain that such correlations exist between molecular properties and various textile parameters such as light fastness, for example. Often these correlations are not evident until calculations on a sufficiently large number of related structures have been performed and the data carefully analyzed. Then appropriate molecular descriptors can more readily be identified and used as input into a neural network.

7 Acknowledgement

The authors would like to acknowledge the US Department of Commerce, National Textile Center (Grant #I98-P01) for financial support of this research.

8 References

1. Kojima M., Degawa M., Hashimoto Y. and Tada M. (1991), *Biochem. Biophy. Res. Commun.*, **179**, p. 817.
2. Hashimoto Y., Degawa M., Watanabe H. K. and Tada M. (1981), *Gann*, **72**, p. 937.
3. Degawa M., Miyairi S. and Hashimoto Y., (1978), *Gann,* **69**, pp. 367.
4. Degawa M., Shoji Y., Masuko K. and Hashimoto Y. (1979), *Cancer Lett.*, **8**, p. 71.
5. Miller J. A. and Miller E. C. (1961), *Cancer Res.*, **21**, p. 1068.
6. Hashimoto Y., Watanabe H.K. and Degawa M., (1981), *Gann,* **72**, p. 921.
7. Freeman H.S., Posey Jr. J.C. and Singh P. (1992), *Dyes and Pigm.*, **20**, p. 279.
8. Degawa M., Kojima M. and Hashimoto Y. (1985), *Mutation Res.*, **152**, p. 125.
9. Lye J., Hink D and Freeman H.S., *Computational Chemistry applied to synthetic dyes.* In: Cisneros G., Cogordan J.A., Castro M., Wang C. and editors (1997), *Computational chemistry and chemical engineering*, Singapore World Scientific Publ.
10. Spartan v.5.0, *Wavefunction Inc.*, 18401 Von Karmen Avenue, Suite 370, Irvine, CA 92612.
11. Perdew J.P. (1986), *Phys. Rev.*, **B33**, p. 8822.
12. Perdew J.P. (1987), *Phys. Rev.*, **B34**, p. 7046.
13. Chung K.T. and Cerniglia C.E. (1992), *Mutation Res.*, **277**, p. 201.
14. Ashby J., Paton D., Lefevre P.A., Styles J.A. and Rose F.L., *Carcinogenesis*, **3**, 1277 (1982).
15. Cunningham A.R., Klopman G. and Rosenkranz H.S. (1998), *Mutation Res.*, **405**, p. 9.
16. Miller J.A., Sapp R.W. and Miller E.C. (1949), *Cancer Res.*, **9**, p. 652.
17. Mueller G.C. and Miller J.A. (1948), *J. Biol. Chem.*, **176**,pp. 535.

18. Miller J.A. and Miller E.C. (1947), *Cancer Res.*, **7**, p. 39.

19. Zadeh L. (1965), Fuzzy Sets, *Information and Control,* **8**, pp. 338-353.

20. Kosko B. (1986), Fuzzy Entropy and Conditioning, *Information Sciences,* **40**, pp. 165-174.

21. DeLuca A. and Termini S. (1972), A Definition of a Nonprobabilistic Entropy in the Setting of Fuzzy Sets Theory, *Information and Control*, **20**, pp. 301-312.

22. Kaufmann A. (1975), Introduction to the Theory of Fuzzy Subsets, *Academic Press,* New York.

23. Klir G.J. and Folger T.A. (1988), *Fuzzy Sets, Uncertainty and Information*, Prentice Hall, Englewood Cliffs.

24. Knopfmacher J. (1975), On Measures of Fuzziness, *J. Math. Anal. and Appl.*, **49**, pp. 529-534.

25. Loo S.G. (1977), Measures of Fuzziness, *Cybernetica*, **20**, pp. 201-210.

26. Yager R.R. (1979), On the Measure of Fuzziness and Negation. Part I: Membership in the Unit Interval, *International Journal of General Systems*, **5**, pp. 221-229.

27. Yager R.R. (1980), On the Measure of Fuzziness and Negation. Part II: Lattices, *Information and Control*, **44**, pp. 236-260.

28. Higashi M. and Klir G.J. (1982), On Measures of Fuzziness and Fuzzy Complements, *International Journal of General Systems,* **8**, pp. 169-180.

29. Kosko B. (1992), *Neural Networks and Fuzzy Systems*, Prentice Hall, Englewood Cliffs.

30. Zadeh L. (1983), A Computational Approach to Fuzzy Quantifiers in Natural Languages, *Comput. Math. Appl.*, **9**, pp. 149-184.

31. Zadeh L. (1983), The Role of Fuzzy Logic in the Management of Uncertainty in Expert Systems, *Fuzzy Sets and Systems*, **11**, pp. 199-227.

32. Dombi J. (1982), A General Class of Fuzzy Operators, the De Morgan Class of Fuzzy Operators and Fuzziness Measures, *Fuzzy Sets and Systems*, **8**, pp. 149-163.

33. Cios K.J. and Sztandera L.M. (1992), Continuous ID3 Algorithm with Fuzzy Entropy Measures, *In: Proceedings of the 1st International Conference on Fuzzy Systems and Neural Networks*, IEEE Press, San Diego, pp. 469- 476.

34. Cios K.J., Goodenday L.S. and Sztandera L.M. (1994), Hybrid Intelligence Systems for Diagnosing Coronary Stenosis, *IEEE Engineering in Medicine and Biology,* **13**, pp. 723-729.

35. Sztandera L.M. (1990), Relative Position Among Fuzzy Subsets of an Image, *M.S. Thesis*, Computer Science and Engineering Department, University of Missouri-Columbia, Columbia, MO.

36. Vedrina M., Markovic S., Medic-Saric M. and Trinajstic N. (1997), *Computers Chem*, **21**, pp. 355-361.

37. Balaban A.T. (1982), *Chem Phys Letters*, **89**, pp. 399-404.

Chapter 4 Neural-Fuzzy Systems for Color Classifications in Textiles

Bugao Xu
Department of Human Ecology
The University of Texas at Austin
Austin, TX78712

Summary: This chapter introduces two applications of neural networks, fuzzy clustering and fuzzy logic to color classifications in textiles. The first application is the identification of color patterns on a printed fabric. Color separation is necessary when the colorfastness of a printed fabric is evaluated because different colors in the fabric may change in different rates during the laundering or other treatments. A regular colorimeter cannot perform color separation and reports only the average color of an area that may contain multiple colors. The self-organizing-map and fuzzy clustering algorithm can be used to automatically separate colored patterns for independent evaluations. The second application in this chapter is the color classification of cotton fibers using fuzzy logic. Cotton colors are grouped into a number of classes having blurring and overlapping boundaries. The partition of the classes in the current grading diagram does not reflect the grouping nature of cotton colors, yielding significant disagreements with the visual ratings. Fuzzy logic appears to be effective in dealing with ambiguity and uncertainty in cotton color grading.

Key words: color identification, self-organizing map, fuzzy clustering ,fuzzy logic.

1 Automatic Color Classification in Printed Fabrics

1.1 Introduction

Colorfastness of dyed or printed fabrics is an important quality attribute indicating the fabric wearability. To evaluate the colorfastness of a fabric, colors on the fabric need to be measured before and after a specific treatment, such as laundering or abrasion. For a solid-color fabric, a colorimeter can be placed anywhere on the fabric to read its surface color. Since a printed fabric often contains multiple colors with intricate patterns, a colorimeter may not be suitable for measuring areas of multiple colors because it outputs only the average color of an area being tested. In order to evaluate multiple colors of a printed fabric, regions with different colors must firstly be separated and color measurements must be taken individually over each region. Color image processing technology provides robust ways for color identification and evaluations. In this chapter, we introduce an automatic color identification algorithm to separate color patterns in printed fabric images. The algorithm involves using an unsupervised neural network to group image pixels into a number of major color clusters, and the fuzzy c-means clustering to further classify pixels ambiguous to any color cluster.

1.2 Self-Organizing Map --Identification of Major Colors

A self-organizing map (SOM) is an unsupervised and non-parametric neural network proposed by Kohonen in early 1980's [1]. It converts patterns of arbitrary dimensionality into the responses of a two–dimensional array of neurons, which is often referred to as a feature map [2,3,4]. One of important characteristics of the SOM is that the feature map preserves neighborhood relations of the input pattern. A typical SOM structure, shown in Figure 1, consists of only two fully-connected layers: one input layer and one output layer. In the image analysis applications, the number of input neurons is equal to the number of the pixels of an input image. Each input neuron is connects to all the output neurons, which are arranged in a planar array.

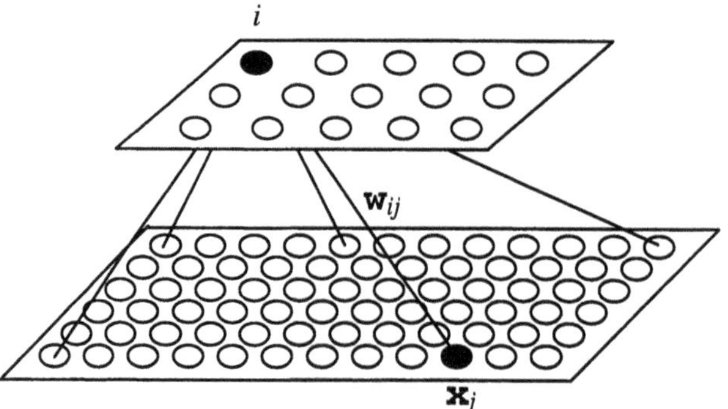

Figure 1 A self-organizing map

Let i denote the ith output neuron and A denote the whole output set (i.e., $i \subset$A). An input element, $x_j=[x_1, x_2, ..., x_p]_j$, is a vector belonging to a p-dimensional Euclidean space $\boldsymbol{R}^p$ ($\mathbf{x}_j \in \boldsymbol{R}^p$, j=1, 2,…, n). The weight vector connecting the jth input neuron and the ith output neuron is $\mathbf{w}_{ij} = [w_1, w_2, ..., w_p]_{ij}$. According to the Kohonen's learning rule, the output neuron c is the winner for the input x_j at time t of learning if

$$\left\|\mathbf{x}_j - \mathbf{w}_{cj}^{(t)}\right\| = \min_{i \in A}\left\{\left\|\mathbf{x}_j - \mathbf{w}_{ij}^{(t)}\right\|\right\}.$$

The winning neuron c and its neighbors are then updated by modifying their weight vectors towards the current input as follows:

$$\mathbf{w}_{ij}^{(t+1)} = \begin{cases} \mathbf{w}_{ij}^{(t)} + \alpha^{(t)}[\mathbf{x}_j - \mathbf{w}_{ij}^{(t+1)}], & \text{for } i \in N_c^{(t)} \\ \mathbf{w}_{ij}^{(t)}, & \text{otherwise} \end{cases}$$

where $N_c^{(t)}$ represents the neighbors of neuron c at time t and α is a learning constant. As a result of weight adjustments, the planar feature map is obtained with weights coding the stationary probability density function of the input vectors used for training. A wide range of the neighborhood kernel $N_c^{(t)}$ is often chosen at the beginning of the learning process to guarantee the global ordering of the weight vectors. Both of its width and height decrease slowly during learning. This learning process leads patterns that are similar in some measure to be aggregated on the feature map. After a SOM is convergent to a balance state, each output neuron on the feature map indicates the occurrence of similar patterns in the input. In an image-analysis application, each output neuron indicates a certain cluster of patterns in the image, and the average of the weight vectors connecting the neuron with the pixels in the set is the center of the cluster. In a cluster, the difference between any pixel and its center does not exceed the given limit with respect to the defined measures, such as color attributes. Since a feature map represents the counts of the pixels with similar measures, it is also called a density map.

To identify color patterns in a 24-bit color image, the color of each pixels, i.e., $x_j = [R_j, G_j, B_j]$, was used as the input vectors of the SOM. The output layer was designed to have 49 neurons organized in a 7x7 matrix. Each output neuron collects the counts of the pixels whose color differences are within a given tolerance. On the density map, some neurons receive more responses from the input than others because different colors are presented by different numbers of pixels. If an image consists of only m different colors, there will be m neurons receiving distinct values in the density map. If an image contains color patterns that have many transitional colors among several major colors, the density map will show multiple peaks and valleys. By connecting all the valleys with straight lines, the density map is partitioned into small regions, each of which corresponds to a major color cluster in the input image. Thus, a problem of identifying major clusters in an image becomes a problem of finding peaks and valleys in the density map. If the major colors have substantially more pixels than the transitional colors, the boundaries among the partitioned regions will be crisp. Figure 2 shows a fabric with four distinct color strips (*a*), and the density map generated by the

SOM (*b*). Since the colors in each stripe are quite uniform, the boundaries are well defined in the density map. Figures 3*a*-*d* exhibits the regions corresponding to the four color clusters in Figure 2*a*.

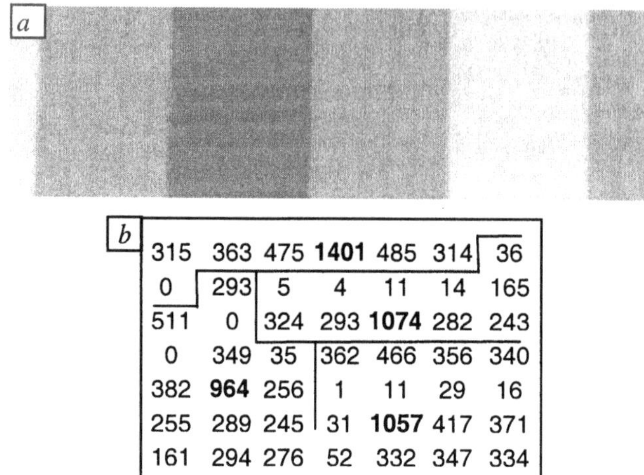

Figure 2 Color identification by SOM, *a*: fabric with color stripes; *b*: density map

Many printed fabrics consist of intricate color patterns that vary in size, yielding more complex peak-valley patterns in their density maps. Some peaks may be adjacent. Figures 4*a* and *b* show a printed fabric and its density map, on which there are four pairs of adjacent peaks indicated by dashed boxes. Whether a pair of peaks should be treated as two separate color clusters or one color cluster depends on how close the centers of the two color clusters. Suppose there are N_p peaks found on a density map and the density values at these peaks are arranged in a descending order:

$$Density\ (i) > Density\ (i+1), \qquad i = 1,2,\ldots, N_p - 1\,.$$

Note that all the N_p peaks in this order. Because the first peak has the maximum density value, it corresponds to a color with the most pixel count in the image, and should be always counted as a major color cluster. The average $\overline{\mathbf{w}}_1$ of all the weight vectors w_{1j} (j=1, 2., ..., n) connecting to this peak neuron is taken as the center of the cluster. The second peak may be assigned to a new color cluster or combined to the first cluster depending on how close the two cluster centers are in the input space.

Assume that **P** is the set of the found peaks, and E is a predefined error tolerance for checking the similarity of any two peaks. If peak k has been identified as a color cluster, peak l will be considered as a new cluster if

$$\min_{l \in \mathbf{P}} \left\| \overline{\mathbf{w}}_k - \overline{\mathbf{w}}_l \right\| > E \,.$$

Once all the **P** peak neurons have been checked, a set of **C** clusters may be found. For the non-peak neurons in the density map, a merging step is taken to combine them with the existing **C** clusters. A non-peak neuron i will be assigned to cluster c, if

$$c = \arg\min_{k \in \mathbf{C}} \left\| \overline{\mathbf{w}}_k - \overline{\mathbf{w}}_i \right\| .$$

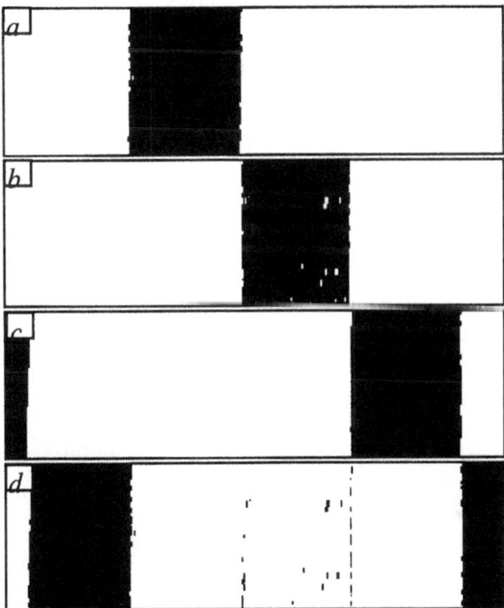

Figure 3 Color identification by SOM, *a-d*: four partitioned regions.

After the merging, the density map is split into **C** regions that correspond to the **C** color clusters. According to the belongingness of each neuron, each pixel in the image can be classified into a cluster to which the pixel has the shortest distance.

In Figure 4b, the bolded numbers are the densities of the nine initially-found peak neurons. After the distance-checking, the nine peaks were reduced to five peaks (see dashed boxes). The other non-peak neurons were further merged with their closest peaks. The straight lines in Figure 3*b* show how the map was split. Figures 4*c-g* are the regions that show the five major colors identified by the SOM. These separated regions enable the calculation of the average color value of all the pixels in each region individually, and a new color image (Figure 4*h*) can be reconstructed for color evaluation by re-filling the regions with these color values. For the example shown in Figure 4*a*, the average color values of the five color regions are:

$$\bar{\mathbf{x}}_1 = [R_1, G_1, B_1] = [13, 1, 89],$$

$$\bar{\mathbf{x}}_2 = [R_2, G_2, B_2] = [56, 1, 98],$$

$$\bar{\mathbf{x}}_3 = [R_3, G_3, B_3] = [70, 155, 176],$$

$$\bar{\mathbf{x}}_4 = [R_4, G_4, B_4] = [106, 1, 84],$$

$$\bar{\mathbf{x}}_5 = [R_5, G_5, B_5] = [17, 119, 153].$$

By comparing this image with the original image in Figure 3*a*, it can be found that there are pixels that are still misclassified. This is because in addition to the five major color clusters, there are pixels whose colors gradually change between multiple clusters. The belongingness of these transitional pixels is ambiguous, and may not be properly classified using the "hard" boundaries. To deal with the uncertainty of transitional colors, it is more advantageous to use a fuzzy clustering method for finer classifications. The clustering results obtained from the SOM can provide useful information for the initialization of a fuzzy clustering.

1.3 Fuzzy c-Means Clustering--Fine Color Classification

Fuzzy clustering is a procedure to partition patterns into a number of relatively homogeneous clusters with additional information supplied by the cluster membership values indicating degrees of belongingness for all the input elements.

The fuzzy c-means (FCM) algorithm is one of the widely used clustering methods developed by Bezdek [5]. For a given set of data x_j (j=1, 2,..., n) and a given number of clusters c, FCM is an algorithm of optimizing an objective function defined as follows:

$$J(u_{ij}, \mathbf{v}_i) = \sum_{i=1}^{c}\sum_{j=1}^{n} u_{ij}^m \left\| \mathbf{x}_j - \mathbf{v}_i \right\|^2, \qquad m > 1,$$

where u_{ij} is the membership value in [0,1] that expresses the degree to which the element x_j belongs to the ith cluster, m is a given exponential weight used to change the degree of fuzziness of the membership values, and v_i is the center vector of cluster i. u_{ij} should satisfy the following constrains:

$$\sum_{i=1}^{c} u_{ij} = 1, \qquad j = 1,2,\dots,n \text{ , and,}$$

$$0 < \sum_{j=1}^{n} u_{ij} < n, \qquad i = 1,2,\ldots,c.$$

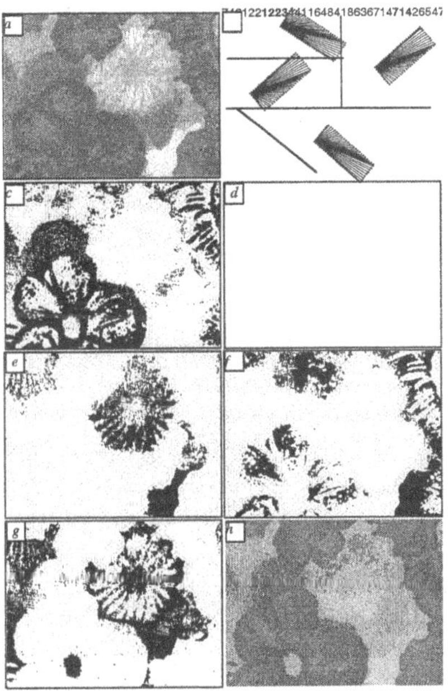

Figure 4 Color identification by SOM, *a*: original image; *b*: density map; *c-g*: five partitioned regions; *h*: reconstructed color image.

The central vectors and the membership values are updated iteratively according to the following two equations:

$$\mathbf{v}_i = \frac{1}{\sum_{j=1}^{n} (u_{ij})^m} \sum_{j=1}^{n} (u_{ij})^m \mathbf{x}_j, \qquad i = 1,2,\ldots,c,$$

$$u_{ij} = \frac{(1/\|\mathbf{x}_j - \mathbf{v}_k\|^2)^{1/m-1}}{\sum_{k=1}^{c} (1/\|\mathbf{x}_j - \mathbf{v}_k\|^2)^{1/m-1}}, \qquad i = 1,2,\ldots,c; j = 1,2,\ldots,n.$$

The minimization of $J(u_{ij}, \mathrm{v}_i)$ leads to an optimal partition of the input data set into c clusters, and the final clustering result is stored in the membership matrix:

$$\mathbf{U} = [u_{ij}]_{i=1,\ldots,c;j=1,\ldots,n}.$$

It can be seen that the equations for calculating u_{ij} and v_i cannot be solved analytically. The FCM algorithm takes an iterative approach to approximate the minimum of $J(u_{ij}, v_i)$ starting from a given position. When the number of clusters c, the exponential weight m and the initial values of the cluster centers $\mathbf{v}_i^{(0)}$ (i=1,2, ..., c) are chosen, the initial membership values $u_{ij}^{(0)}$ (i=1,2, ..., c, j=1,2, ..., n) can be calculated.

From $u_{ij}^{(0)}$, the next iteration $\mathbf{v}_i^{(1)}$ and $u_{ij}^{(1)}$ can be proceeded. The iteration continues until the maximum difference of the membership values in two consecutive iterations is smaller than the give termination criterion ε, i.e.,

$$\max_{i,j} | u_{ij}^{(l+1)} - u_{ij}^{(l)} | < \varepsilon.$$

The FCM algorithm does not deal with how to determine the appropriate number of clusters, c, and the initial values of $\mathbf{v}_i$ (i=1,2,...,c), both of which can greatly affect the clustering results. In order to make the color identification process automatic and objective, we use the number of clusters and the average color value of each partitioned cluster as the initial position of the FCM, i.e.,

$$\mathbf{v}_i^{(0)} = \overline{\mathbf{x}}_i, \ (i=1,2,...,c).$$

In fact, it proves that this way of selecting the initial parameters helps the FCM algorithm achieve better and faster clustering than a random selection. The m and ε values are often selected as 2 and 0.01, respectively.

Figure 5 presents the FCM clustering results of the same sample used in the SOM (Figure 4*a*). The number of the clusters was set to five, and the initial center $\mathbf{v}_i^{(0)}$ of each cluster was assigned to the average color value of the cluster identified by the SOM. After 25 iterations of the FCM clustering, many pixels were reassigned to different clusters, forming five new color clusters shown in Figures 5*a-e*. These regions are noticeably different from those in Figures 4*c-g*, and the average color values of the new clusters are:

$$\mathbf{v}_1^{(25)} = [5, 2, 79],$$
$$\mathbf{v}_2^{(25)} = [35, 4, 105],$$
$$\mathbf{v}_3^{(25)} = [141, 181, 188],$$
$$\mathbf{v}_4^{(25)} = [103, 1, 84],$$
$$\mathbf{v}_5^{(25)} = [30, 132, 163].$$

In a colorfastness test, the images of a multi-color fabric sample can be captured before and after the sample treatment (laundering or abrasion), and then regions with different colors can be separated using the procedure described above. After the color separation, color changes can be measured for each individual region.

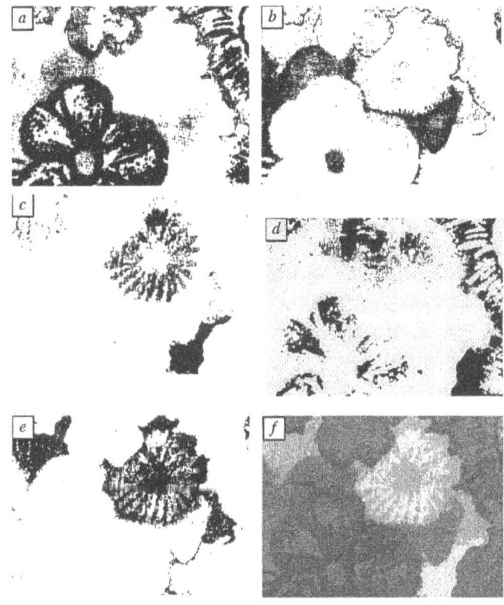

Figure 5 Color clustering by FCM, *a-e*: five partitioned regions; *f*: reconstructed color image.

1.4 Conclusion

Multiple color patterns in a printed fabric can be automatically identified using a hybrid method of self-organizing map and fuzzy c-means clustering. A SOM converts a color image to a planar density map which indicates the pixel counts of each major color cluster, and is designed primarily for determining the number of major color clusters in the image and the average color value of each cluster. Based on the results from the SOM, the FCM algorithm is used to further classify colors of pixels by solving the uncertainty of transitional colors with membership values. This two-step classification method results in an objective separation of color regions in an image and therefore enables color evaluations on an individual basis.

2 Cotton Color Classification by Fuzzy Logic

2.1 Introduction

The Nickerson-Hunter color diagram illustrates a partition of a two-dimensional color space (R_d, b) that defines the USDA color grades, and has been used in the colorimeter of high-volume-instruments (HVI) to grade cotton colors [9]. The color diagram consists of one set of linear lines nearly in the horizontal directions and another set of non-linear lines nearly in the vertical directions, forming the boundaries of various color grades. During the past several decades, the HVI color grading was not accepted by the industry, because it did not achieve a satisfactory agreement with visual grading [10,11]. One of the major reasons attributing to the disagreement is that some of the color boundaries in the diagram, especially the one between white and light spotted color classes, do not properly separate neighboring color classes. This may be due to the fact that the diagram was established more than 30 years ago and cotton fiber colors had chronic shifts because of changes in seed varieties and environments. Another main reason for the disagreement is that the crisp, abrupt separations of color grades in the diagram do not reflect the clustering nature of cotton color classes, which often have blur boundaries. Neighboring classes always overlap to some extent. Thus, the belongingness of a sample point in an overlapping region is inherently ambiguous.

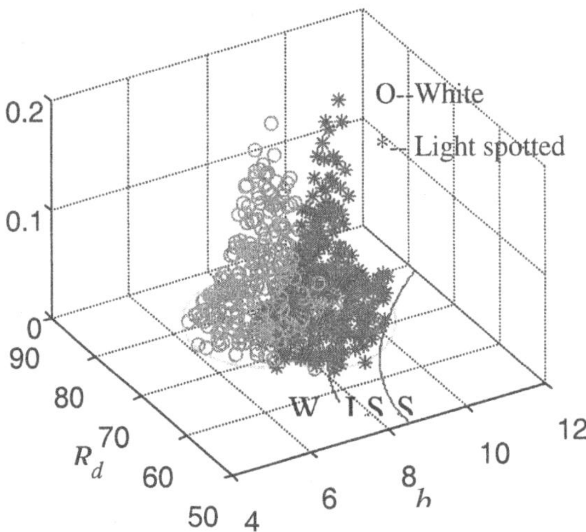

Figure 6 Distributions of white and light spotted colors classified by classers.

The above analysis is evidenced by the color data of 2,489 bales of cotton selected from the 1996 U.S. crop. To facilitate the discussion, we focus on the analysis of two major color classes, white and light spotted, which are the two most disputable color grades [10]. Figure 6 shows the distributions of the white and light spotted classes labeled by classers (o-- white, *-- light spotted). Both the white and light spotted classes seem to follow a two-dimensional Gaussian distribution. Note that the real boundaries separating three major color classes, white (W), light spotted (LS) and spotted (S), were also drawn on the R_d-b plane. Although the two classes have distinct populations, they overlap extensively and their intersection does not seem to coincide with the W-LS boundary. It is evident that the W-LS boundary does not provide a realistic separation between the white and light spotted classes. This mismatch brings a systematic error into the HVI's color grading.

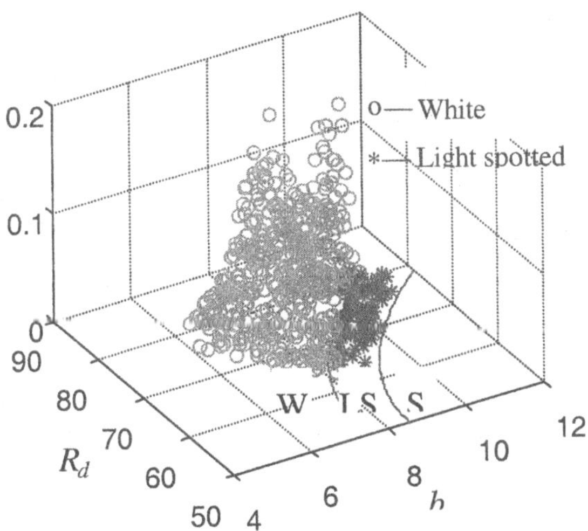

Figure 7 Distributions of White and Light Spotted Colors Classified by the HVI

In Figure 7, the distributions of the white and light spotted classes classified by the HVI is presented. The clear split between the white and light spotted classes rises from the crisp boundary used by the HVI. However, the W-LS separation by the HVI does not indicate the natural grouping of the cotton color data in these two classes. It is logic to consider that the two peaks of the distribution represent two separate populations in the color data as seen in Figure 6. But the HVI did not allocate these two populations properly. This is the reason why the HVI tends to grade cotton colors for the white class more likely than for the light spotted class.

In order to make the machine grading more realistically reflect the natural grouping of cotton colors, the Agriculture Marketing Service of the USDA and the cotton community agreed to adjust the boundaries of the Nickerson-Hunter color diagram. This measure effectively reduced the systematic bias in the color grading with the HVI colorimeters, and therefore officially adopted as official grading starting from year 2000. However, the modified color diagram does not deal with problems associated with blur, overlapping boundaries of color classes. In a previous paper, we presented how to use an artificial neural network to reduce the machine-classer disagreements [10]. The neural network acts as a black-box classifier that does not use explicitly defined boundaries. In this chapter, we will report an investigative work of applying fuzzy logic to eliminating the hard boundary problems in cotton color grading.

Fuzzy logic uses the fuzzy set theory and approximate reasoning to deal with imprecision and ambiguity in decision-making [6-8]. It provides intuitive, flexible ways to create fuzzy inference systems for solving complex control and classification problems. For classification applications, fuzzy logic is a process of mapping an input space into an output space using membership functions and linguistically specified rules. In this study, we take the output of the HVI colorimeter, the R_d-b data, as the input, and color grades as the output. Our discussion will be limited to the classification for five major color classes, white (W), light spotted (LS), spotted (S), tinged (T) and yellow stained (YS). Figure 8 presents a schematic diagram of the fuzzy inference system (FIS) for cotton color grading [7].

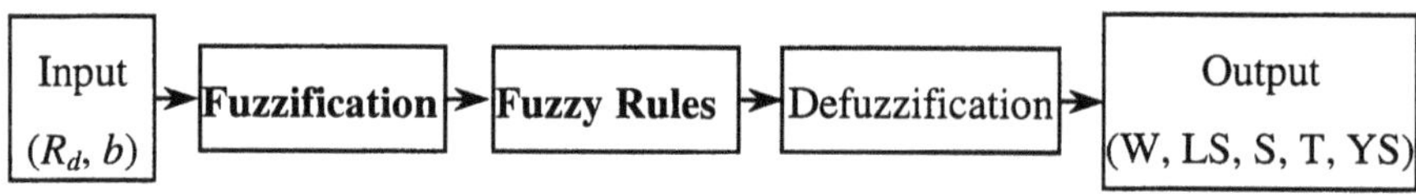

Figure 8 Fuzzy Inference System for Cotton Color Grading

2.2 Fuzzy Sets and Membership Functions

Elements in ordinary or crisp sets have full memberships in one set and zero memberships in others. A fuzzy set contains elements only with partial memberships ranges from 0 to 1 to describe uncertainty for classes that do not have sharply defined boundaries. For each input and output variable of an FIS, fuzzy sets are created by dividing its universe of discourse (entire space) into a number of sub-regions and are named in linguistic terms. Fuzzy sets' linguistic terms are useful in establishing fuzzy rules. In designing an FIS for cotton color grading, five fuzzy sets were selected for the input variable R_d and six for b. The fuzzy sets for R_d represent five levels of brightness varying from very low (I), low (II), median (III), high (IV) to very high (V), and the fuzzy sets for b represent six levels of yellowness ranging from very low (I) to extremely high (VI). Table I presents the ranges and other distributions parameters of the input fuzzy sets. Each fuzzy set overlaps with its adjacent fuzzy sets. The reason for adding one more fuzzy set for b is that b seems more critical than R_d in determining cotton major color classes (white, light-spotted, etc.). In general, the more intermediate levels are used, the higher accuracy the classification would be. But increasing the fuzzy sets will significantly increase the number of fuzzy rules in the next step. The final selection on the number of fuzzy sets and their range may be determined by trial and error. Since this FIS was designed to classify five major color classes, the output variable was split by five fuzzy sets named as white (1), light-spotted (2), spotted (3), tinged (4) and yellow stained (5). The range of the output variable was equally divided into five sections for the five fuzzy sets.

Table 1 Parameters for the Fuzzy Sets of the Input Variables

Fuzzy set	R_d			b		
	Range	m	σ	Range	m	σ
Very low (I)	40-52.5	46.5	3.5	4-7	4.0	1.00
Low (II)	45-65	55	3	4-11	7.2	1.00
Medium (III)	54-75	64	3	7-12.5	9.5	0.80
High (IV)	60-82.5	71.5	3.5	9-17	12.4	1.19
Very high (V)	67.5 87.5	77.5	3.5	11-18	15.1	1.19
Extremely high (VI)				14-18	18.0	1.19

Once the fuzzy sets are chosen, a membership function for each set should be created. A membership function is a curve that maps an input element to a value between 0 and 1 showing its degree of belongingness to a fuzzy set. The curve can have different shapes, such as bell (Gaussian), sigmoid, triangle and trapezoid, for different types of fuzzy sets [6,7]. In this study, the Gaussian distribution curve was used to build the membership functions for the input fuzzy sets R_d and b:

$$\mu(x) = e^{-(x-m)^2/2\sigma^2}$$

where m and σ are the mean and the standard variation of one fuzzy set in x (R_d or b). Finding the right parameters for the functions is a major task, which may be selected arbitrarily and then tweaked by using a known set of input-output data. The m and σ values used in this FIS are included in Table 1, and the membership functions are displayed in Figure 9. The extent of overlap between the membership functions of two adjacent sets indicates the nature of the unsharp boundary between two color classes.

For the simplicity of defuzzification, a triangular shape was used to construct the membership functions for the output fuzzy sets:

$$\mu(x)=\begin{cases}=0, & x<a \text{ or } x\geq c\\ =(x-a)/(b-a), & a\leq x<b\\ =(c-x)/(c-b), & b\leq x<c\end{cases}$$

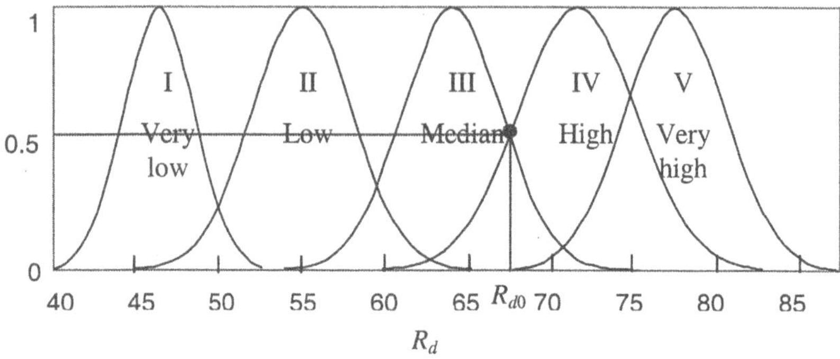

Figure 9 Membership functions of input fuzzy sets

The shape and size of the triangular function depend on the values of *a*, *b* and *c*. To make the output clear and unbiased, the symmetric, non-overlapping and equal-size membership functions were used for all the output sets (Figure 10).

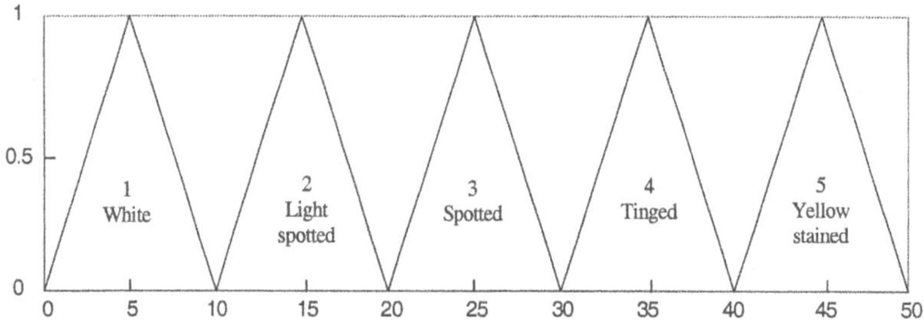

Figure 10. Triangular membership functions of output fuzzy sets

2.3 Fuzzification

Fuzzification is a step to determine the degree to which an input data belongs to each of the appropriate fuzzy sets via the membership functions. For a given input point (R_{d0}, b_0), the memberships of all the fuzzy sets are calculated, and only the fuzzy sets with non-zero memberships are forwarded to the next step. In Figure 9, an example of determining the relevant fuzzy sets was shown for an input data (R_{d0}, b_0)=(67.5, 9.0). R_{d0} belongs to the medium (III) and high (IV) sets of R_d with the memberships being 0.51 and 0.52, while b_0 belongs to the low (II) and medium (III) sets of b with the memberships being 0.20 and 0.82. There are four combinations with the selected fuzzy sets:

$$[\mu_{\mathrm{III}}(R_{d0}), \mu_{\mathrm{II}}(b_0)] = [0.51, 0.20],$$
$$[\mu_{\mathrm{III}}(R_{d0}), \mu_{\mathrm{III}}(b_0)] = [0.51, 0.82],$$
$$[\mu_{\mathrm{IV}}(R_{d0}), \mu_{\mathrm{II}}(b_0)] = [0.52, 0.20],$$
$$[\mu_{\mathrm{IV}}(R_{d0}), \mu_{\mathrm{III}}(b_0)] = [0.52, 0.82].$$

These four combinations will be evaluated by fuzzy rules to determine the output fuzzy sets and the weight of each rule influencing the output.

2.4 Fuzzy Rules

In an FIS, fuzzy rules provide qualitative reasoning that links input fuzzy sets with output fuzzy sets. They are a collection of linguistic rules of the form [8]:

R_i: **If** $\underline{R_d \text{ is } A_i}$ AND $\underline{b \text{ is } B_i}$, **then** $\underline{\textit{color} \text{ is } C_i}$, i=1, 2, …, k

where A_i, B_i and C_i are the fuzzy sets for the inputs R_d and b and the output color in the ith rule R_i, and k is the number of the rules. The values of A_i and B_i are the linguistic terms such as very low (I) and very high (V), and the values of C_i are the linguistic terms such as white (1) and light spotted (2). An example of such a rule may be given as follows:

If $\underline{R_d \text{ is } \textit{very high} \text{ (V)}}$ AND $\underline{b \text{ is } \textit{very low} \text{ (I)}}$, **then** $\underline{\textit{color} \text{ is } \textit{white} \text{ (1)}}$.

The **if**-part of the rule is called the antecedent, and the **then**-part of the rule is called the consequent. Since the antecedent in this FIS always involves two conditions (one for R_d and one for b), fuzzy operators are needed to specify the relationships of the fuzzy sets in the antecedent. AND (intersection), OR (union) and NOT (complement) are the three common fuzzy operators. Because R_d and b should be simultaneously observed in selecting a color class, the fuzzy operator in all the antecedents must be AND. For two fuzzy sets A and B, the fuzzy AND is defined as [6,8]:

$$A \text{ AND } B: \min\{\mu_A(x), \mu_B(x)\}.$$

Fuzzy AND aggregates two membership functions by outputting the minimum value at a given input x (R_d or b). The result of fuzzy AND serves as a weight showing the influence of this rule on the fuzzy set in the consequent.

The fuzzy rules should be established based on both visual grading experience and the basic relationships between color data and color grades in the HVI color diagram. Since there are five fuzzy sets in R_d and six fuzzy sets in b, there are 30 possible combinations in the antecedents when only fuzzy AND is applied. Thus, the maximum number of the fuzzy rules that can be established is 30. To simplify the fuzzy rule expressions, we designed a chart that illustrates (Figure 11). In the chart, the five sets of R_d is arranged vertically and the six sets of b is arranged horizontally. A knot (a black dot) between a horizontal line and a vertical line indicates an antecedent formed by the connecting fuzzy sets from R_d and b, and leads to a consequent that in turn gives an output fuzzy set. The aggregated membership of the antecedent is then used as a weight factor to modify the size and shape of the membership function of the output fuzzy set in a way of either truncation or scaling. Truncation is done by chopping-off the triangular output function, while scaling is done by compressing the function.

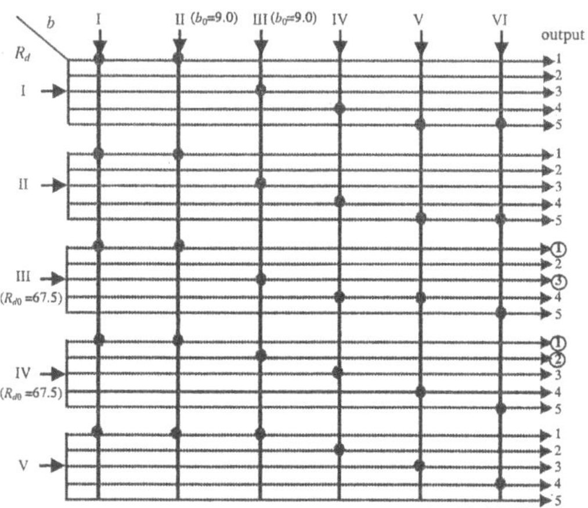

Figure 11. Fuzzy Rules

In this FIS, the truncation operation was used. If the membership function of an output fuzzy set is $\mu(x)$ and the weight generated from the antecedent is w, the truncated functions are:

$$\mu^T(x) = max\{\mu(x), w\}.$$

As given previously, the input fuzzy sets containing sample (R_{d0}, b_0) have four combinations, which satisfy the four following rules:

R_1: **If** <u>R_d is *median* (III)</u> AND <u>b is *low* (II)</u>, **then** <u>*color* is *white* (1)</u>;

R_2: **If** <u>R_d is *median* (III)</u> AND <u>b is *median* (III)</u>, **then** <u>*color* is *spotted* (3)</u>;

R_3: **If** <u>R_d is *high* (IV)</u> AND <u>b is *low* (II)</u>, **then** <u>*color* is *white* (1)</u>;

R_4: **If** <u>R_d is *median* (IV)</u> AND <u>b is *median* (II)</u>, **then** <u>*color* is *light spotted* (2)</u>.

The four output fuzzy sets were circled in Figure 11. The weights of the rules on the four outputs are 0.20, 0.51, 0.20 and 0.52, respectively. The truncated membership functions of the output fuzzy sets were presented in Figure 12.

2.5 Defuzzification

After all the fuzzy rule evaluations are done, the FIS needs to output a crisp member to represent the classification result (color classes) for the input data. This step is called defuzzification. As seen in the (R_{d0}, b_0) example, one input data may generate several weighted output fuzzy sets. The multiple sets need to be aggregated into a single set in preparation for the defuzzification. If those output fuzzy sets are different, the aggregation can be done simply by placing all the truncated functions together to form the final fuzzy set. If two of the output fuzzy sets are identical, they can be combined by using fuzzy OR, which is defined as [6,8]:

A OR B: $max\{\mu_A(x), \mu_B(x)\}$.

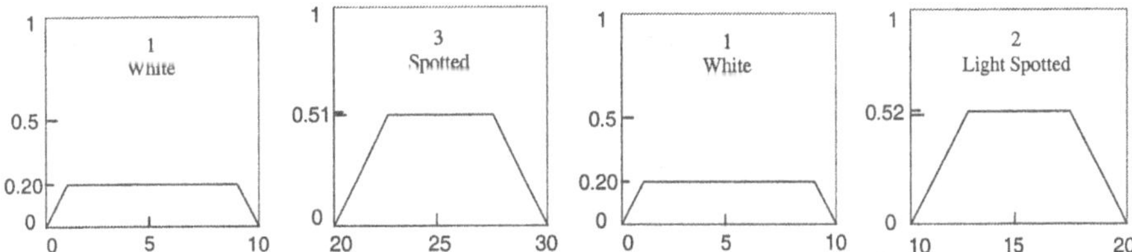

Figure 12 The weighted membership functions of output fuzzy sets

Fuzzy OR gives the maximum value of the two membership functions at any given point. For example, sample (R_{d0}, b_0) has two 'white', one 'light spotted' and one spotted output functions. After the fuzzy OR operation, the two trapezoidal 'white' functions merge into one so that the aggregated curve becomes the one shown in Figure 13.

The most popular method for defuzzification is the centroid calculation, which returns a grade weighted by the areas under the aggregated output functions. Let a_1, a_2, ... a_n be the areas of the truncated triangular areas under the aggregated function, and c_1, c_2, ... c_n be the coordinates of their centers on the x-axis. The centroid of the aggregated area is given by [4,7]:

$$G = \sum_{i=1}^{n} a_i c_i \Big/ \sum_{i=1}^{n} a_i$$

The location of the centroid indicates the color class to be designated to the input data. For sample (R_{d0}, b_0), G is 17.1, which falls in the light spotted set (see

Figure 13). The location of this centroid inside the identified fuzzy set can suggest a finer rating.

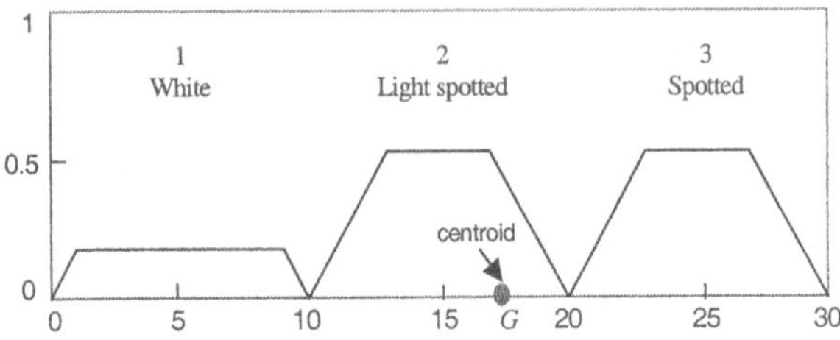

Figure 13 Aggregation of the weighted output functions

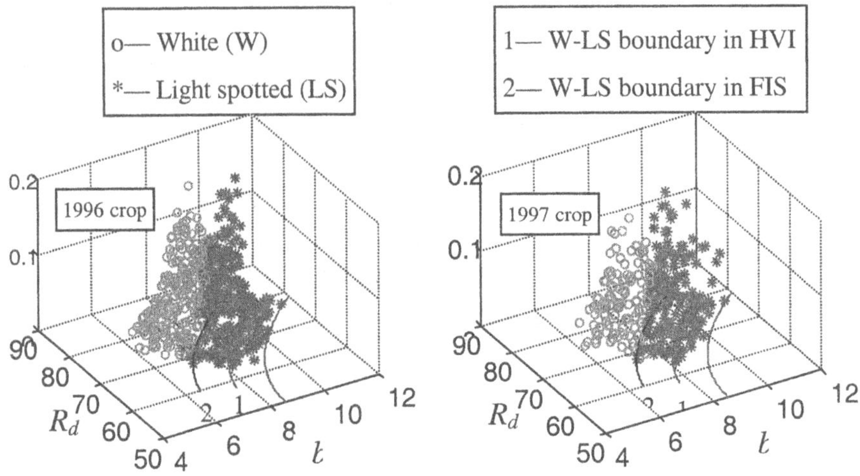

Figure 14 Distributions of White and Light Spotted Colors Classified by the FIS

2.6 Experiment

The constructed FIS was tested using cotton samples in 1996, 1997 and 1998 crop years. Since the majority of the U.S. Upland cottons are 'white' and 'light spotted' and the most disputable grading occurs between these two classes, samples only in these two classes were selected for the experiment. There were totally 2489 samples from the 1996 crop, 1375 samples from the 1997 crop and 658 samples from the 1998 crop in the selection. The samples were firstly graded by official cotton classers, and then measured by the HVI colorimeters. The (R_d, b)

data were converted into color grades using the HVI color diagram and the FIS. We used the classers' grades as a reference to check the consistency of the HVI and FIS data, since the classers' grades were the official grades at that time and they were coincided with the natural grouping of the color data. Table 2 presents the data showing the disagreements among the testing methods. % disagreement was calculated by dividing the disagreement count with the number of the tested samples. For the white class, 54.1% of the samples in 1996 crop, 61.5% in 1997, and 51.2% in 1998's were graded differently by the classers and the HVI. The amount of disagreements between the classers and the FIS has decreased considerably, and the FIS results are consistent for consecutive years. The classer-FIS disagreement much more evenly spread between the white and light spotted classes, suggesting that there is no systematic bias in the FIS's grading.

Table 2 Disagreements Between Different Grading Methods

Color Class	White			Light Spotted		
Crop Year	1996	1997	1998	1996	1997	1998
By Classer	1240 (49.8%)	416 (30.3%)	464 (70.5%)	1249 (50.2%)	959 (69.7%)	194 (29.5%)
By HVI	2350 (94.4%)	1200 (87.3%)	628 (95.4%)	139 (5.6%)	175 (12.7%)	30 (4.6%)
By FIS	1161 (46.6%)	355 (25.8%)	447 (68.0%)	1328 (53.4%)	1020 (74.2%)	211 (32.0%)
C-HVI Disagreement	1347 (54.1%)	846 (61.5%)	337 (51.2%)	9 (0.4%)	14 (1.0%)	3 (0.5%)
C-FIS Disagreement	154 (6.2%)	87 (6.2%)	40 (6.1%)	234 (9.4%)	144 (10.5%)	72 (10.9%)

Figure 14 shows the distributions of the white and light spotted color classes of the 1996 and 1997 samples classified by the FIS. It can be seen that the two color classes in both years' data were reasonably separated. Based on the two newly separated data clusters, a best-fit polynomial curve was calculated (curve 2 in the Figure). This curve provides a new boundary that more properly reflects the segregation between the white and light spotted color classes. The distribution for the 1998 crop was not included because the number of the available samples from that year was not sufficient for drawing an effective distribution.

2.7 Conclusion

This section presents the investigative work of using fuzzy logic to construct a fuzzy inference system for classifying the white and light spotted cottons based on their (R_d, b) color measurements. The performance of the FIS depends primarily on the selections of input and output fuzzy sets, the design of the membership functions and the establishment of fuzzy rules that guide the input-output relationships. These parameters may be initially selected based on an analysis of a small set of data, and then tweaked by trial and error with more data. The classers' knowledge on color grading can be incorporated into the fuzzy rules to enhance the overall agreement with visual grading. The FIS is able to perform consistent classifications for the samples from different years. Although the generalization of the FIS could not be tested more thoroughly in this study due to the lack of samples and other resources, the preliminary results has shown great potential for being a more reliable way of grading cotton colors.

3 References

1. Kohonen, T. (1990), The Self-Organizing Map, Proc. of IEEE, Vol.**78**, No.90, pp1464-1480.

2. Zhang, X. and Li, Y.(1993), Self-Organizing Map as a New Method for Clustering and Data Analysis, Proc. of IJCNN'93-NAGOYA, Japan, pp2448-2451.

3. Kaski, S.(1997), Data Exploration Using Self-Organizing Maps, Ph.D thesis, Neural Networks Research Center, Helsinki University of Technology, Finland.

4. Lin, C.T. and Lee, G. (1996), Neural Fuzzy Systems – A Neuro-Fuzzy Synergism to Intelligent System, Prentice-Hall Inc.

5. Bezdek, J.C. (1981), Pattern Recognition with Fuzzy Function Algorithms, Plenum Press, New York.

6. Cox, E. (1999), Fuzzy systems Handbook, 2nd Edition, Academic Press, Chestnut Hill, MA.

7. Jang, Roger J. S. and Gulley, N. (1997) MATLAB/Fuzzy Logic Toolbox, MathWorks, Inc. Natick, MA.

8. Hguyen, H.T. and Walker, E.A. (1999), A First Course in Fuzzy Logic, 2nd Edition, Chapman & Hall/CRC, Boca Raton, FL.

9. United States Department of Agriculture (1993), The Classification of Cotton," Agricultural Handbook.

10. Xu, B., Su, J., Dale, D. and Watson, M.D. (2000), Cotton Color Grading by Neural Network, *Text. Res. J.*, **70**, 430-436.

11. Xu, B., Fang, C. and Watson, M.D. (1998), Investigating New Factors in Cotton Color Grading, *Text. Res. J.*, **68**, 779-787.

Chapter 5 Agent-Based Modeling of the Textile/Apparel Marketplace

E.L. Brannon, S. Thommesen, and T. Marshall

Department of Consumer Affairs

Auburn University

Auburn, AL 36849 USA

1 Agent-Based Modeling as a Method of Inquiry

A number of approaches exist where populations of interacting agents are linked in a network of connections and studied using simulation. The problem is that such models tend to settle into a self-consistent pattern that fails to provide information about the constant adaptation present in real world systems. Agent-based modeling derives from a basic insight: a system of agents following simple rules can exhibit complex behaviors. At each time step, each agent assesses its own condition in comparison to its preference set, evaluates the environment within its sight range, and determines its behavior in the next time step. The collective actions of all agents influence the developing pattern of behavior and the decision environment at each successive time step. Such systems are termed co-evolutionary models because agents must adapt to each other and to changes in their world.

Unlike other modeling approaches, no agent operates with perfect global knowledge. Instead agents in the simulation operate in their world with incomplete knowledge using heuristics rather than pure rationality, much like consumers and firms in the real world. Thus, agents are not able to optimize their fitness or utility. The resulting models exhibit perpetual novelty where agents adapt to niches of opportunity as they occur.

Agent-based modeling has been used to simulate ecosystems, markets, and social systems-——non-equilibrium systems in which no coordinating authority dictates the system's behavior. Advances in computer processing speed, memory, and graphics in the last decade have increases opportunities to study adaptive agents through simulation.

Science writers surveying pioneering efforts in the field of agent-based modeling explained the approach as an alternative to the traditional linear, reductionist approach to inquiry (Levy, 1992; Waldrop, 1992; Holland, 1995). As a method of inquiry agent-based modeling can be useful for theory building, generating hypotheses for empirical testing, and for hypothesis testing. The method is particularly valuable for phenomena where agents' actions occur over a long time span, the system is inaccessible for observation, or the dynamics of a system are subtle. This method of inquiry can even be used to recreate an historical system and to play what-if games involving alternative scenarios.

Researchers use agent-based modeling to study complex, self-organizing systems. After the simulation is designed and programmed, it becomes a laboratory for rigorous, replicable computer-based experiments. A simulation run presents an opportunity to observe agents' actions across time, identify emerging patterns of behavior, and interpret findings. By using the same set of variables but resetting initial conditions, investigators explore different scenarios. As understanding of the system evolves, researchers can adjust model parameters or revise rule sets to enable the study of ever more complex interactions.

This chapter summarizes a program of research to investigate the feasibility of agent-based modeling to simulate aspects of the textile/apparel marketplace. The challenge facing the research team was to:

- Develop prototype simulations using agent-based models that demonstrated characteristics of adaptive agents.
- Use simulation of agent-based models as a computer laboratory--performing computer experiments, observing the emerging behaviors of agents, and interpreting results within theoretical frameworks.
- Determine the viability of using these simulations to strategize about evolving markets and as a tool for executive decision support.
- Evaluate agent-based modeling as a method of inquiry for building theory, generating hypotheses, and testing hypotheses.

The simulations discussed in this article resulted from collaboration between academic researchers, programmers, software developers (initially at the Santa Fe Institute, later at SWARM Development), and industry executives managing change in a non-equilibrium environment. The National Textile Center, which provided funding for the project, actively supports such team-based efforts through its research consortium of government agencies, universities, and the apparel industry.

1.1 A Brief History of the Field

Agent-based modeling became widely popular in scientific circles as a result of the rise of interest in the field of "Artificial Life" (A-life) (Langton, 1989.) A-life research is based on the modeling of autonomous agents which collect information about their environment (including other agents) and interact with it. These agents contain algorithms that allow them to evolve and/or adapt to their environment (Gutowitz, 1995). Agent rule sets may be deterministic or probabilistic, as the case requires, and may change as the agents learn or evolve. From an A-Life perspective:

- The essence of life is in the process; the medium (physical structure, body) in which it occurs is incidental.
- Self organization arises out of complexity ("life at the edge of chaos"), as a result of natural laws.

A-life research has two very different, but complementary strands:

- "Weak" A-life where the goal is to use agent-based computer modeling (simulation) as a tool to investigate real-world systems both in the physical and the social sciences, in order to "enhance the descriptive and predictive understanding of actual and potential life processes" (Tesfatsion, 1997).
- "Strong" A-life where the goal is to synthesize within a computer, a robot, or some other medium, software entities that can be said to possess at least some of the qualities usually associated with being alive.

The program of research reported in this chapter fits within the "weak" A-life framework.

Agent-based modeling has been used to simulate ecosystems, markets, and social systems. A list of some early A-Life models illustrates the great diversity of interests that can be accommodated within the field:

- ECHO, a platform for classifier system modeling evolution through reproduction and mutation (Holland, 1975; Holland, 1995).
- TIERRA, an environment for the study of the evolution of populations of simple computer algorithms (Ray, 1992).

- CELLSIM, a general-purpose cellular-automaton simulation toolkit (Langton, 1986).
- MANTA, an agent-oriented ant simulator (Drogoul, 1993).
- Biomorphs, a model of evolving animal shapes (Dawkins, 1989).
- BOIDS, a model showing the flocking behavior of birds as arising solely out of simple, local behavior rules (Reynolds, 1987).
- TRANSIMS, a simulation of traffic patterns in the city of Albuquerque to investigate traffic jams and pollution patterns analysis (Nagel, et al., 1996)[1].

Agent-based modeling has been used to investigate animal interactions (e.g. of ants, birds, etc.), interactions of inanimate pieces of matter (e.g. molecules in a gas or in a liquid, genetic molecules, etc.), and human systems. The mode of analysis adopted in A-life studies is called "bottom-up" as contrasted with the "top-down" approach of the usual analytical models. What this means is that the model explicitly attempts to model the behavior of the lowest-level entities of interest at a realistic level of complexity and specificity. This means that each agent's knowledge is *limited*, and that its rules for interaction with other agents are *local* (depending on the states and actions of a limited number of other units in its immediate vicinity, suitably defined). The complexity of the system, rather than being built into each agent, arises instead from the interactions among the units. That is, the more complex order *emerges* from the simple behaviors of the agents. In modeling a human system, a bottom-up approach would be to study how coordination might arise (emerge) as a result of the interactions of less-than-perfectly rational agents, as opposed to a top-down approach which *imposes* order by fiat through the assumption of perfect agent rationality and the fictional device of a directing entity.

In addition to its roots in the field of A-life, agent-based models also share some common features with a widely used modeling technique in the field of Operations Research (OR), Discrete-Event Simulation. Both model a set of entities or agents which act independently according to a set of rules, and both contain a scheduling mechanism that keeps track of which agent needs to perform which action at which point in simulation time. But A-life and discrete-event simulation have very different perspectives on their subject matter. A-Life researchers are looking for qualitative patterns in macro behavior, especially emergent behaviors not deliberately programmed into agents at the micro level. Discrete Event simulations, on the other hand, look for quantitative answers to system behaviors,

1 For a more complete description of these and other A-life projects, see Langton, et al., 1992; Langton, 1994; Brooks and Maes, 1995; Adami, 1998.

using small numbers of carefully specified micro entities whose behaviors are known. The goal is to find some optimal configuration of the system being modeled. In that case, the focus of interest is not in emergent behaviors (in fact, it may be desirable to rule such behaviors out). The program of research reported on in this chapter draws on the experience of both traditions, but is closer to the A-life perspective.

1.2 Steps in Designing and Programming a Simulation

Agent-based modeling is a top-down approach in the design phase because components of the simulation (organization, events to model, distributions, probabilities) are derived from empirical research and theorizing about antecedents, events, and subsequent conditions. In the implementation phase agent-based modeling is a bottom-up method of inquiry because behavior emerges during simulation that cannot be predicted solely from knowledge of its constituent parts. In the case of simulation of the textile/apparel marketplace, agent-based modeling held the promise that interpretation of emergent behavior in the simulation would lead to new strategies and tactics for executives and new questions for empirical researchers.

Creating simulations of agent-based models requires deep understanding of the inputs, outputs, and behaviors within a system—a theory of the system. Derived from either observation and previous research or reverse engineering the system, the theory is translated into precise specifications—parameters and variables, agents' actions and preferences, rule sets for the agents and their environment, and timetables for actions and updating agents. The programming of the simulation is not a trivial process because decisions about how to display data, the stages of initializing the simulation, and selection of a suitable pseudo-random number generator can influence simulation validity. Competence in research methods is required to verify and validate the simulation and to use it as a platform for experimentation.

1.2.1 The Software Platform for Agent-Based Modeling

Robert Axtell and Joshua Epstein at the Brookings Institution had a simulator they called Sugarscape, programmed in Objective Pascal for the Macintosh. This simulator was used to carry out the research reported on in Axtell and Epstein (1996). The authors' goal was to grow artificial societies "*in silico*" in order to discover local and micro mechanisms sufficient to generate social structures and collective behaviors. They used simulation to investigate cultural processes, trade, and disease transmission. (A CD containing animated video clips showing the running of some of their models is available from the authors at the Brookings Institution web site http://www.brook.edu.) However, the authors did not make the source code for this simulator available to other researchers.

The fledgling Swarm simulation package offered a more accessible portal to agent-based modeling. Swarm was conceived as a "virtual computer" for managing concurrent interactions of large numbers of agent-type objects. In the mid-1990s it was being developed under the direction of Chris Langton, then at the Los Alamos National Laboratory, as a tool for investigating questions in the field of Artificial Life. It was conceived from the start as an Open Source project, which means that users get access to the full source code for the package. By attending lectures and seminar series organized by the Santa Fe Institute, the authors of this paper were introduced to the software and became early beta testers of the program. Soon the program became available to a wider range of users and an international network of researchers in physical, biological, and social sciences began using Swarm to model systems from highway traffic to market dynamics.

Swarm consists of a set of program libraries written in Objective-C. These provide support for the basic functions needed to put together an agent-based modeling simulation: scheduling, random-number generation, graphical display of agents and objects, and other aspect of models. Users write their simulations in either Objective-C or in Java, and support is being added for additional languages[2].

Researchers beginning in agent-based modeling today can choose from a large number of simulation packages. Two such Open Source packages are close to Swarm in scope and capabilities.

- ASCAPE, a rewrite of the Sugarscape simulator in Java, is available from the project web site at http://www.brook.edu/es/dynamics/models/ascape. [For a more complete description of ASCAPE, see Parker, 2001.]
- RePast, a simulator developed by Nick Collier and others at Social Science Research Computing at the University of Chicago is very similar to Swarm but written completely in Java and is available from the project web site at http://repast.sourceforge.net.

Like Swarm, ASCAPE and RePast are general-purpose agent modeling tools. While making the modeler's life much easier since they provide all the pieces necessary to create a simulation of an agent-based model, these simulators still require the user to be a reasonably proficient programmer, and this places a hurdle in the way of the average person wishing to explore agent-based modeling.

2 For more detailed descriptions of the Swarm package see Burkhart (1994) and Terna (1998). The Swarm website (http://www.swarm.org) contains the latest version of the software, documentation, sample Swarm simulations, and links to a large number of projects using Swarm. The Swarm-support mailing list links beginning Swarm users and more skilled users and includes lively exchanges on agent-based modeling and programming simulations.

Programs which have tried to make agent-based modeling more accessible to those without extensive programming skills include:

- StarLogo, created at the MIT Media Lab and available from the web site http://lcs.www.media.mit.edu/groups/el/Projects/starlogo.
- AgentSheets, a commercial package available from the web site http://www.agentsheets.com.

StarLogo is a specialization of the "turtle" based Logo programming language. It is designed to make it especially easy for students to explore populations of interacting agents driven by simple rules. Using StarLogo students gain insight about the emergence of macro-level complexity from simple micro-level behaviors. [For a more complete description of StarLogo and several simulations undertaken using this package, see Resnick (1994).] AgentSheets tries to insulate the user from the programming step by use of a graphical model description, which is automatically turned into lower-level Java code.

1.2.2 Reliability and Validity Issues

Once a prototype of the simulation is complete, the next stage is verification and validation. In verifying the simulation, investigators run trials of the model exploring combinations of parameters and variables. Verification determines if the simulation design captures the theory of the system being studied and if the internal logic of the model is operating in a consistent and reasonable way. Verification may also include evaluation of the model by experts in the field. In the validation stage, results from the model are matched against existing data. Problems at this stage lead to revision of the system specifications.

If the purpose of the project is exploratory—for building theory, generating hypotheses for empirical testing, or strategizing—this level of development may be sufficient to produce a useful simulation, one that simplifies a phenomenon so that it can be understood and studied while retaining real-world veracity. Theory building may be the most immediate use of simulations of agent-based modeling because investigators oscillate between specifying theoretical parameters and variables and exploring the behavior of the model under various scenarios in an inductive approach to science (Seror, 1994).

If the purpose of a project is hypothesis testing or prediction, then a model that is plausible against existing data and persuasive with experts is not sufficient. In addition, the model must be calibrated. That is, model parameters must be given plausible values taken from the system or population to be modeled. Then test the calibration by verifying that the model yields correct or plausible results under specified conditions.

For methodological completeness, investigators must establish experimental validity and perform sensitivity analysis (Gilbert & Doran, 1994). Experimental validation means designing experiments that fully explore a wide range of combinations of parameters and variables to adequately sample the model's behavior. Unlike laboratory experiments where standard procedures have been established, procedures for computer experiments are still being developed. Questions remain about how many time steps and how many runs of the simulation represent adequate sampling, what data should be captured and how, and what documentation is sufficient for replication by other scientists.

Interpretation of results from computer experiments can be problematical, especially if findings are counterintuitive, but sensitivity analysis can bolster confidence in the model. In a sensitivity analysis, the investigator examines assumptions underlying the model to identify their relative significance to the functioning of the simulation. While a simulation may not be able to prove specific theorems the way a mathematical model can, they may be used to disprove hypotheses (Axtell, 2000).

2 Simulation for the Textile/Apparel Marketplace

Marketplace dynamics have been identified as a fertile arena for agent-based modeling (Anderson, Arrow, & Pines, 1988). A system is complex when a great many independent agents (consumers, fashion firms, media) interact with each other in a great many ways. The richness of these interactions allows the system to undergo spontaneous self-organization. The system itself is adaptive to events, actively trying to turn them to some advantage. This is an apt description of the textile/apparel marketplace because it constantly forms and reforms with shifts in tastes, lifestyles, immigration, technological developments, prices of raw materials, and myriad other factors. Such a system can be viewed as a complex, adaptive, self-organizing system—that is, the actions of agents (consumers, firms, industries, economies) create a system in dynamic equilibrium. Agents behaving in the marketplace are not completely rational with easy predictability. Instead, they think and act on the basis of expectations and strategies. Through the agents' mutual accommodation and mutual rivalry, new structures and behavior patterns emerge continuously (Holland & Miller, 1991; Waldrop, 1992). The idea that complex behaviors result from relatively simple sets of rules offers a construct for examining and modeling consumer behavior and the competitive marketplace.

As agents form new structures, those structures acquire collective properties. Buyers and sellers behave like agents in a complex system because they interact with each other sharing information in a social transaction called word-of-mouth. Through self-selection some buyers cluster around certain products or styles forming patterns of preference. Consider, for example, the collective properties of style tribes where individuals form a group based on shared viewpoint, lifestyle, and way of dressing (Polhemus, 1996). Or, consider the collective properties of partnerships between retailers and manufacturers in licensing agreements to produce private-label merchandise. Co-evolution on each level inevitably leads to the emergence of new structures that engage in new behaviors. The trend is always to more and more complex behaviors. It is this worldview that helps executives in the fashion world manage high velocity change in the next millennium.

Simulations of marketplace systems offer industry executives a window through which to observe these systems in action under various sets of parameter and variable settings. Agent-based models can be used to:

- Gain insights about marketplace dynamics based on the emergent behavior in systems of adaptive agents.
- Revitalize fundamental theoretical frameworks that guide industry decision making and joint industry/academic research programs.
- Leverage expertise and accelerate understanding of marketplace phenomena by involving industry executives and the academic community in building and testing simulations.

Simulations allow executives and researchers to view patterns in data and introduce a richer environment for theory building and strategizing about evolving markets. The effect is similar to flight simulators—a way to get the feel for how situations develop and important variables interact. Such simulations will not predict the outcome of a new acquisition or what new fashion will emerge next season, but they will provide statistical and structural measures of the process and guidance in answering business questions like: When do you introduce a new product? How big an impact will it cause? How does the introduction of a new product affect existing products? How do you recognize when a good has become central to a system as opposed to a fad? What effect will increased marketing support have on adoption of an innovative product? How does product mix relate to marketplace alliances? Which interactions between consumer preferences and a proposed product will lead to a purchase decision? Using simulation facilitates the investigation of these and other similar questions because it makes the interactions more transparent, more easily observed, and more accessible for experimentation.

The research team has demonstrated the potential of agent-based modeling through the design and programming of three simulation prototypes:

- InfoSUMERS, a simulation of diffusion of innovation.
- Sphere of Influence, a simulation of consumer/supplier interactions.
- Virtual Consumer, a simulation of the formation of purchase intent.

3 InfoSUMERS: A Diffusion of Innovation Simulation

3.1 Theory of the System

Rogers (1962,1995) proposed a diffusion of innovation model in the form of a bell-shaped curve. The left side of the curve represented the earliest adopters of innovation, the center section majority adoption, and the right side late adoption by laggards. The model launched innumerable studies of the demographics and psychographics of consumers and of the way innovation moves through a population. Rogers' concepts are important because today's marketing managers use these ideas as a guide for the dissemination of new technologies, products, and services.

Rogers (1962) original model shows a very small group of innovators who begin the diffusion process followed by a larger group of opinion leaders. Innovators are first to adopt a new fashion and attend more to mass media sources as an information source (Polegato & Wall, 1980). Opinion leaders make it visible within their social groups through the power of personal influence. Innovators were expected to make up 2.5 percent of the total adopters, early adopters/opinion leaders added an additional 13.5 percent. Fashion followers include both the majority adopters who swell the diffusion curve to its highest point and those who adopt after that. After the peak, the number of new adopters decreases. Rogers' model served as the theoretical framework for studies of the demographics and psychographics of innovators and opinion leaders, not only in the field of fashion but in all product categories and many kinds of social systems.

In Rogers' view of diffusion of innovation, an innovation enters the system where it may disappear almost immediately or agents may adopt it and spread news of the innovation to others. In fashion terms, the innovation may be the invention of a new fiber, a new finish for denim, introduction of an unusual color range, a modification in a silhouette or detail, a different way to wear an accessory, or a mood expressed in a distinctive style. Once introduced, it diffuses through the population as more and more consumers have a chance to either accept or reject it. This pattern of acceptance or rejection determines the innovation's lifecycle. The diffusion process maps the response to the innovation over time.

On occasion adoption of an innovation reaches a critical mass and sets off a chain reaction of decisions resulting in a "hit" (Farrell, 1998). What appears to be many people simultaneously coming to the same decision, really started with a single innovation that built in popularity as networks formed and patterns of interaction emerged. None of this happened because some outside force was coordinating the action.

Yet by the late 1980s research on diffusion within consumer social systems had reached a state of malaise because of the emphasis on direct relationships (main effects) (Gatignon & Robertson, 1985). Critical elements like competitive activity, segmentation strategy, and marketing mix decisions were rarely part of the conceptualization and empirical research on diffusion. Identification and targeting of consumer innovators and opinion leaders for product promotion had proven difficult, partly because an individual may be an early adopter in one product category and a laggard in another (Behling, 1992). Large-scale diffusion as represented by the fashion system is difficult to study using traditional empirical methods because innovation begins from many sources and diffuses through different mechanisms to different consumer segments (Crane, 1999). Simulation of agent-based models offers a new approach for studying a system that can't be described completely using mathematical models because it allows investigators to examine structural and dynamic properties while testing the dependence of results on parameter selection and assumptions (Axtell, 2000).

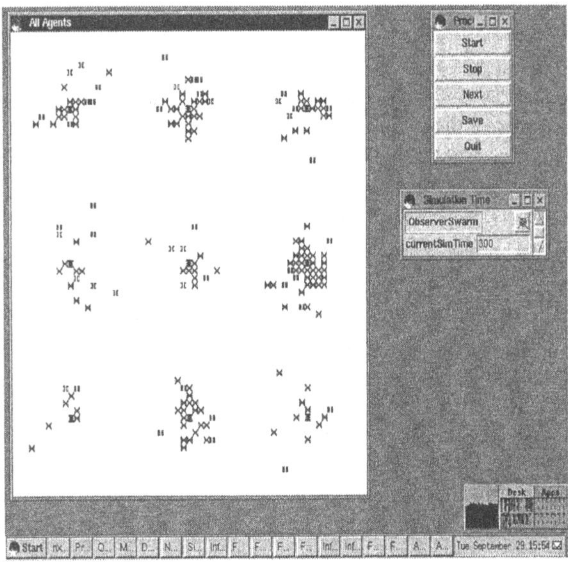

Figure 1: Agent interaction pattern formed after 300 time steps using InfoSUMERS simulation.

3.2 Simulation Design

Diffusion of innovation is the theoretical framework of the simulation InfoSUMERS (designed by E. Brannon and S. Thommesen and programmed in Swarm software). InfoSUMERS models the process whereby fashion information is transmitted from industry sources to consumers who use the information to decide which "look" to adopt and promote to others. The current InfoSUMERS prototype is the second version. The first version was evaluated against the theory of diffusion of innovation and extensively revised. The revision provided more complex rule sets for agents and a redesigned data collection scheme.

InfoSUMERS creates a world inhabited by adaptive agents of three types:

- Influence Agents (IA) representing fashion's gatekeepers (e.g., manufacturers, media, and merchants) who select from the looks proposed those that get promoted and disseminated and stimulate change through Impersonal Influence.
- Fashion Leaders (FL) representing individuals who are attuned to fashion trends, open to change, monitor the cues of Influence Agents and other Fashion Leaders, and influence others though Personal Influence.
- Fashion Followers (FF) representing individuals who take cues from Fashion Leaders and each other and includes individuals with a range of decision profiles.

Agents are randomly generated as individuals with decision profiles consisting of the following parameters:

- The **Strength of the Influence** message transmitted by the agent --a level that can be distributed uniformly or normally within set limits over the population.
- The agent's **Pattern of Output** for influence--as a streak at either high or low levels for a certain number of time periods.
- The **Change Threshold**--the amount of accumulated influence needed for an agent to 'change its mind'—a level that can be distributed over the population in a specified way.
- The **'Blackout' Period**, a time after an agent has changed to a 'new look' during which an agent outputs influence promoting that new look and does not accumulate influence about other looks.

Thus, each agent is an individual with goals and preferences that determine actions.

The currency in the InfoSUMERS world is influence—agents collect and release influence into the environment where it both diffuses and evaporates. The influence is tied to a particular fashion look—the agent releasing influence promotes its current look. Other agents in the neighborhood either collect and accumulate the influence or not according to their own goals and preferences. At each time step, each agent evaluates the situation and determines its action according to its individual rule set. Possible actions include doing nothing, collecting influence from the environment, changing looks, and releasing influence into the environment. The rule sets include probability functions so that each decision is unique and adaptive.

Initiation of the simulation includes setting the rate of diffusion and evaporation, the number of each kind of agents, and the number and kind of fashion looks available for adoption, and the source of the innovation. During each run of the simulation, data are gathered at each time step. A graph shows the number of agents changing looks for the first time. A bar chart shows the penetration of each look by agent type. Screen capture is used to record spatial patterns of agents in the world at selected time intervals (usually every 100 time steps). The experiment ends when convergence occurs—that is, when a single look dominates choice across all agents.

3.3 Computer Experiments

Sociologist Georg Simmel wrote at the turn of the last century about fashion, locating the phenomenon in the individual's dual motivation to fit in and stand out (1904). As Simmel saw it, the dual drives of imitation and differentiation can never be satisfied and the perpetual oscillation drives the engine of fashion change. In this single essay, Simmel identified the motivation for participating in fashion, the arena where these motivations play out, and speculated about the factors influencing the speed of change.

Benvenuto (2000) reviewed the Simmel essay and pointed out that simulation in the social sciences tends to focus on the imitative aspects. He suggested that a simulation that captures the tension between imitation and differentiation would make a contribution to the field since Simmel's ideas can be applied to phenomena in other fields of behavior.

Because InfoSUMERS was designed around the concept of cultural transmission of fashion information, it offered a laboratory to investigate Simmel's speculations about fashion change. Specifically, this paper reports on a set of experiments investigating Simmel's contention that "the more nervous the age, the more rapidly its fashions change". By setting the parameters for diffusion and evaporation in the simulation and the rule sets for three categories of agents—Influence Agents, Fashion Leaders, and Fashion Followers—it was possible to investigate the speed of fashion change under two conditions of "nervousness":

- **Agent Nervousness** where a low Change Threshold and a short Blackout Period modeled the nervous agent who is quick to change looks.

- **Market Nervousness** where high levels of diffusion and evaporation represented a nervous market where influence is released but quickly dissipates.

The Table shows the average number of time steps for convergence under various scenarios. Results show that the more "nervous" the agents (low Change Threshold, short Blackout Period), the fewer time steps to convergence—a finding that dovetails with Simmel's prediction. The "nervous" agent has a lower resistance to change. Therefore, all the agents in the simulation to varying degrees are acting like innovators and early adopters. If innovations were added to the simulation after initial convergence, another quick convergence would occur. In such a fashion environment, innovations would be in constant demand and would cycle through the system at a rapid pace.

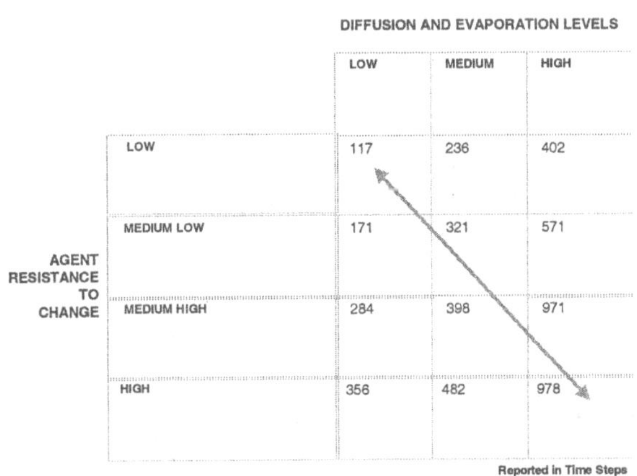

AGENT RESISTANCE TO CHANGE	DIFFUSION AND EVAPORATION LEVELS: LOW	MEDIUM	HIGH
LOW	117	236	402
MEDIUM LOW	171	321	571
MEDIUM HIGH	284	398	971
HIGH	356	482	978

Reported in Time Steps

Figure 2: Time Steps to Convergence Under Scenarios Suggested by Simmel

Results show that the more "nervous" the market (high rates of evaporation and diffusion), the more time steps to convergence—a finding that seems to contradict Simmel's assertion that "the more nervous the age, the more rapidly its fashions change". Where influence burns off rapidly, agents are denied enough time to accumulate influence that would lead to a change in looks. If new looks were added, convergence would not be enhanced. In the "nervous" market scenario, the innovation is not passed from agents like innovators and early adopters (Fashion Leaders in the simulation) to early and late majority and laggards (Fashion Followers in the simulation) in an orderly way. Thus, many innovations are wasted because they enter the system and disappear without making much impact. In such a fashion environment, the rate of fashion change would be slow no matter how many innovations were introduced.

Using InfoSUMERS, the authors were able to design a computer experiment to test an hypothesis about how the fashion system functioned proposed by Simmel (1904). The simulation allowed the researchers to explore the specific conditions under which this theoretical assertion may operate—part of the process of theory building. The results suggest additional avenues for empirical research (e.g., historical studies of periods that match the scenarios, consumer research on "style tribes" where behavior is similar to the "nervous agent"). Industry executives examining the results may see similarities between today's fashion marketplace and the scenarios. Additional simulation experiments could help these executives formulate strategies and tactics for the introduction and support of innovation within the textile/apparel marketplace. Thus, this computer experiment using the InfoSUMERS simulation points to the advantages of this mode of inquiry in the study of cultural transmission and change—for building theory, generating new empirical studies, and strategizing about marketplace dynamics.

4 Virtual Consumer: Simulation of the Formation of Purchase Intent

The purpose of this project was to synthesize existing knowledge about how purchase intention is formed and use that synthesis to design a simulation of the tradeoff analysis leading to the formation of the purchase decision. When fully implemented, Virtual Consumer would provide executives with a computer laboratory for experimenting with prototype products, detailed preference profiles, and scenarios based on proprietary data. While such a simulation will not replace consumer research, it will provide decision support for executives in product development and marketing.

Virtual Consumer is novel from the point of view of simulation and of research methods. To date, agent-based modeling simulated complex social systems as swarms of agents (consumers, companies, economies). Virtual Consumer extends the concept to a single composite agent forming a purchase intention from a "swarm" of interacting sub-agents representing the different facets of the purchase decision. In this project, agent-based modeling is used to simulate a single entity (the apparel consumer) performing a multi-level process, each level composed of a large number of variables with complex interactions within and between levels. Because the simulation model is based on literature-derived components, it becomes a testing ground for theoretical completeness and a source of empirical questions.

4.1 Theory of the System

Consumers review a proposed purchase and confidently state "it's me" or "it's not me". In that instant, the consumer simplifies a cluttered, crowded marketplace using pattern matching and heuristics to determine which garment to consider for purchase and which to eliminate from further consideration. Within that decision space the consumer blends intrinsic and extrinsic motivation, cognition, and affect. The person's selection facilitates both personal and social agendas (Hamilton, 1988). Appearance management is the inclusive concept for the thought processes, activities, and assessment of social implications associated with selecting, buying, and using appearance altering products (Kaiser, 1990). Appearance management includes two system (Brannon, 1993):

- The Self-Presentation System which subsumes all thoughts and feelings about the self and the translation of those thoughts and feelings into strategies of concealment and revelation.
- The Impression Management System which subsumes all thought and feelings about the social implications of dress and the translation of those thoughts and feelings into selection strategies suited to the situation and the desired effect on social interaction.

Each system is further partitioned by preference criteria of two types:

- **Subjective Liking** where acquisition is motivated by symbolic, hedonic, or aesthetic rationales.
- **Reasons-Why** where acquisition is motivated by attribute-based, utilitarian rationales.

Agent-based modeling provides a new mode for investigating the tradeoffs between the self-presentation system and the impression management system and for the role played by subjective liking and more attribute-based considerations in purchase decisions. By making the connections between apparel products; consumers' attitudes, beliefs, and preferences; and constraints on purchase decisions more transparent, the simulation creates a laboratory where researchers can observe decision formation under various conditions.

4.2 Simulation Design

The 'virtual consumer' simulation has three stages: Initiation, Decision Modules, and Diagnostic. In the initiation stage, executives supply available information in the Product Profile and Consumer Profile. The Consumer Profile includes variables for motivation, demographics, lifestyle and income cluster, temperament type, and physical profile. The Consumer Profile may be based on proprietary profiles of current or targeted customers. Any variables not specified by the executive will be set automatically to a default value drawn from consumer behavior research or determined as a probability within a theoretical range. The Product Profile includes variables for price, style genre, fashionability, fabric, fit, season, and consumer adoption factors such as compatibility, complexity, and relative advantage. Values not specified by executives are set to default or probability settings. The simulation sets the sub-agent priorities and message-passing schedule based on these input profiles.

The Consumer Preference Profile is generated from the Consumer Profile based on research findings and likely preference clusters. The Consumer Preference Profile includes variables for fashion preferences (styles, fabric, color, brandedness), fashion orientation, social and fashion anxiety, adopter category, and others.

Tradeoff analysis between the Product Profile and the Consumer Preference Profile takes place in three Decision Modules: Closet Inventory, Risk Evaluation, Price/Value. If there is very little connection between the Consumer Preference Profile and the Product Profile in any module, the simulation run terminates in a "veto" of the purchase. If connections are made between the Consumer Preference Profile and the Product Profile, then the module passes on a "weak buy" or "strong buy" message that moves the decision into the next module.

The **Closet Inventory Module** performs a hypothetical needs assessment to determine the level of urgency to make a purchases decision. The needs assessment sets weights within the Consumer Preference Profile that increases the influence of certain variables. In effect, the Closet Inventory "resets" variables in the Consumer Preference Profile and identifies likely interactions and connections with the Product Profile. Results are reported as a preliminary purchase decision and the weights are retained for subsequent decision modules.

The **Risk Evaluation Module** considers social, enjoyment, and economic risks. Economic risk includes the chance that the purchase price may reduce the ability to buy other products, that the price will fall after purchase, or that other contingencies may devalue the purchase. Enjoyment risk takes into account the potential to become bored by the purchase or of not liking it as much as expected. Social risk involves the fear that the consumer's social group will not approve. Virtual Consumer models these risks under the following conditions:

- Products purchased for either public or private use.
- Products considered either necessities or luxuries.
- Products that are branded or unbranded.
- Products classified as symbolic, hedonic, or functional products.
- The influence of social and reference groups on purchasing.

The Risk Evaluation Module is the most complicated tradeoff analysis because so many interactions are involved. For example, the purchase of a value-expressive product partly to illustrate status in a public arena will include interactions with brand (preference for a "power" brand is more likely such as one with high consumer recognition) and with reference group (anticipation of the opinion of work associates and aspirational Figures will have more importance than those of family members). On the other hand, the purchase of a utilitarian product for private use would have a completely different web of interactions.

The **Price/Value Module** evaluates tradeoffs between consumer wants and needs and product attributes given economic constraints. Price alone is not sufficient to veto the purchase decision if it falls within an theoretical range of possibility and there are overriding factors like weighting for need, the consumer is the type prone to impulsive shopping or with a taste for fashion-forward looks, or the product makes multiple connections with preferences.

Programming for each module is comprised of:

- Rules (variables) that can be set as discrete or continuous values.
- Coordinated rules sets (modules within modules) to show multi-level considerations.
- Connections established between rules sets for message passing on preliminary purchase decisions.
- A weighting system that can be set to favor one rule set or module over all others in the tradeoff analysis.
- Constraint filters that either pass a preliminary purchase decision to the next rule set, modify the decision, or halt the decision (veto the purchase).

At each stage of the tradeoff analysis, the more the proposed product "connects" with the consumer's preferences, the more points accrue leading to a strong buy signal or a weak buy signal. The tradeoff process between competing facets of a complex, multi-layered decision process proceeds until a decision has been reached--a strong buy signal, a weak buy signal, or a not-buy signal. At the end of the simulation a Diagnostic Module details connections (or lack of connections) between the initial Product Profile and Consumer Profile that was used to initialize the simulation run. The Diagnostic Module discloses barriers to the formation of purchase intent and highlights specific tradeoffs that led to the purchase/no purchase decision.

After an initial run, the executive can reset variables in the Product Profile and Consumer Profile to explore what changes—in the product specification or in the identification of the target consumer—could affect the outcome. In this way, the apparel executive uses the simulation to investigate ways to facilitate a positive purchase decision. However, what may seem like an obvious refinement can lead to unexpected outcomes because of emergent properties in the simulation. In this way, the Virtual Consumer simulation provides insight into the complexity of purchase decision formation.

4.3 Virtual Consumer Simulation in Decision Support

The goal in simulating the apparel purchase decision was to provide apparel executives with an individualizable "virtual consumer" for research. When fully implemented, executives can use the Virtual Consumer to experiment with scenarios based on long-tern and short-term company strategies. Simulation runs can be used to identify barriers to the formation of purchase intent under various 'what-if' scenarios. Experimentation with Virtual Consumer can aid the executive in identifying and interpreting the interactions between consumer type and product type resulting in modifications to the product offered or the way the product is marketed. Virtual Consumer also helps apparel executives zero in on specific consumer information needed to produce a more fine-grained profiling of consumers and their purchase behavior to provide more detailed specifications for the model and for subsequent research using the simulation.

Textile/apparel industry executives reacting to presentations on Virtual Consumer have suggested possible applications including:

- Use the simulation to identify the best potential target audience for a new or novel product.
- Set the simulation in batch mode to run a set of products against a set of consumer profiles to identify previously untapped market potential.
- Use the simulation in e-commerce as part of the effort to personalize and customize products for customers.
- Modify the simulation to analyze specific niches like consumers of specific ethnic backgrounds.
- Modify the simulation to evaluate the match between products and their potential in selected international markets.

5 Sphere of Influence: A Simulation of Supplier-Consumer Relationships

5.1 Theory of the System

Sphere of Influence (designed by T. Marshall and N. Terase) is a simulation of the interaction between consumer agents with various preference sets and firms organized around product mix. It allows researchers to investigate brand management strategy given consumers with different preference profiles. Sphere of Influence provides a "big picture" look at a competitive situation, helps users understand why things are happening as they are, and provokes insights on ways to change the dynamics of the situation.

Because the behavioral theory of the firm bridges economics and organizational theory, it provides the theoretical framework. Cyert (1988) was the prime mover in developing the behavioral theory of the firm with a tradition that combined field-based observation and computer simulation as a way to study markets and organizations. The behavioral theory of the firm focuses on the connection between human behavior and activities of the firm. Behavioral models concern themselves with the process of decision making including planning, executing, and analyzing the firm's performance. In this view, people in the firm operate a set of decision rules that tend to be heuristic rather than optimizing. These rules develop as a response to a given environment and evolve from previous decisions. The firm's strategy evolves from the interaction between the firm's actions and the market. Issues important in behavioral models include how organizational objectives are developed, how strategies evolve, and how decisions are reached. Cyert grounded the research program on empirical observations of both the decision processes and the decisions and tested the resulting models in the field with actual organizations, in the laboratory with human subjects, and using computer simulations (Day & Sunder, 1996).

In Sphere of Influence both classes of agents—consumers and firms—are spontaneous and adaptive because there is no external director organizing their actions. However, these two classes of agents represent two kinds of order because decisions within a firm are bounded by a goal such as to gain market share while agents in the market are not bound by common goals. This difference is so fundamental that it is observable in nonhuman systems such as when biologists differentiate between the organization of organisms and ecosystems (Khalil, 1999). Agent-based models like Sphere of Influence offer the opportunity to study both kinds of order as they interact within the simulation.

Exploring scenarios using the simulation can help the people in a firm explore the decision space by making the competitive situation more transparent and more available for experimental manipulation. Thus, the simulation acts as a catalyst for strategizing about current actions and future outcomes—an approach that was an original component of the behavioral theory of the firm.

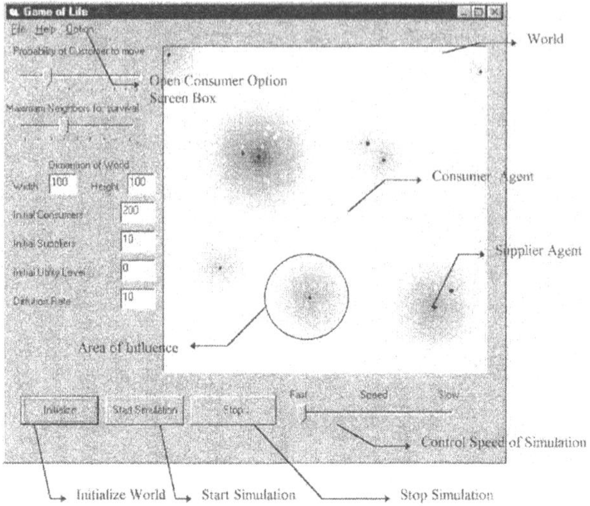

Figure 3: Interaction between Consumer agents and Firm agents in Sphere of Influence simulation.

5.2 Simulation Design

Sphere of Influence uses two agent types: the Consumer and the Firm. The Consumer agent has a overall preference for either a General product mix, for Product A, or for Product B and a specific preference level expressed as a numerical value. The Firm agent either supplies a General product mix, specializes in Product A , or specializes in Product B. In setting up the simulation, the researcher has control of:

- Variables related to the Consumer agent including number of agents, type of agents, preference profile (programmed as a range between minimum and maximum levels), and the probability that a consumer will initiate a move.
- Variables related to the Firm agent including number of agents, type of agents, level of influence exerted by the firm, and preference profile for target consumer (programmed as a range between minimum and maximum levels).

The simulation also has a setting for the diffusion rate of influence within the world that can be manipulated by researchers.

In initiating the simulation, researchers set the Consumer and Firm variables and the diffusion rate to explore some possible scenarios. Setting the Consumer agents' preference profile out of sync with the firms' target customer creates a test of the firms' ability to capture certain consumer segments. If some firms are also given the ability to learn from previous time steps, these firms may, over time, adjust their preference profile for target customers to more closely match that of the Consumer agents. If some firms have this ability and others don't, this creates a test between firms that are consumer-driven and those that aren't.

Once initialized, the Firm agents put out a sphere of influence that attracts some consumers and repels others. When a consumer agent reaches a supplier's sphere of influence, it uses a utility distribution curve and a preference distribution curve to determine whether to move toward or away from the supplier. The resulting assessment can produce a clear decision to move toward or away from the supplier, but if the decision falls into a fuzzy area of indecision, then determination of the behavior for the next time step is dependent on probability. During the simulation run, data are gathered on the level of satisfaction among consumers and the number of consumers by type under the influence of Firm agents.

Sphere of Influence has the potential to investigate more highly developed scenarios. One option allows firms to form alliances in order to extend their influence and capture more Consumer agents. Another option allows Consumer agents to increase their probability of moving based on the failure of firms to match their preferences in previous time steps.

Because it was developed within the theoretical framework of the behavioral theory of the firm, it becomes a laboratory for investigating the decision-making process and the impact of decisions on the agents and the market. As a simulated world of consumer-firm interactions, Sphere of Influence provides a tool for strategizing about the marketplace.

5.3 Sphere of Influence in Decision Support

Sphere of Influence was designed specifically to investigate simulation of agent-based models as a change agent within a strategic environment. For textile/apparel executives the simulation could be used to consider different brand management strategies such as lifestyle brands (products from apparel to furniture and bedding linked by styles offered under one brand), brand tiers (a single company controls multiple brands, each occupying a price and styling position), and niche brands that focus narrowly on a highly desirable target audience. A simulation like Sphere of Influence compresses time and space so that managers see the impact of short-term decisions on long-term outcomes. Executives or college students could use the simulation for presenting cases on strategic initiatives as a way to increase involvement with the task, improve group dynamics, and bolster case performance. By providing a framework for exploring cases and presenting findings, Sphere of Influence enables executives to explore the decision space around a competitive situation and enhances their ability to engage in strategic thinking.

6 Relevance of Agent-Based Modeling to the Textile-Apparel Industry

In recent years the textile/apparel industry structure has shifted away from a "push" system where producers controlled the supply chain to a "pull" system centered on response to consumers. The first generation of changes centered on reengineering manufacturing processes, cycle times, and replenishment systems to reduce costs and increase profits. The next generation of changes centers on defining and segmenting markets down to the individual level and anticipating, developing, delivering, and promoting products to satisfy consumers' needs, wants, and aspirations. The three simulations developed in this research program provide a window on understanding the dynamics of the "pull" system.

Each of the three simulation prototypes developed in this program of research uses a different lens and investigates a different aspect textile/apparel marketplace. By making the connections and interactions more transparent, the simulations provide a platform for investigation, analysis, and strategizing. These three simulations dovetail with structural changes in the industry aimed at perfecting the supplier/consumer relationship by increasing responsiveness to consumers. When the approach is more completely developed, it may be possible to build a "flight simulator" for the exploration of rare or novel marketplace phenomena.

Companies committed to the "pull" system who would be likely to benefit from exploration of agent-based modeling have the following profile:

- Commitment to developing and marketing products to specific consumer profiles.
- Established research systems tracking, analyzing, and evaluating consumer targets.
- Ability to customize and personalize marketing information and products.

7 Summary

The purpose of this program of research was to investigate the feasibility of using agent-based modeling to simulate aspects of the textile/apparel marketplace. The challenge facing the research team was to:

- Develop prototype simulations using agent-based models that demonstrated characteristics of adaptive agents.
- Use this form of simulation as a computer laboratory--performing computer experiments, observing the emerging behaviors of agents, and interpreting results within theoretical frameworks.
- Determine the viability of using the simulation to strategizing about evolving markets as an aspect of executive decision support.
- Evaluate agent-based modeling as a method of inquiry for theory development, generating hypotheses, and hypotheses testing.

Three prototype simulations were developed, each looking at the textile/apparel marketplace through a different lens.

- InfoSUMERS modeled the social system where adaptive agents trade and accumulate information and determine future actions based on their individual preference profiles.
- The Virtual Consumer simulation modeled the complex tradeoffs in decision making within a single agent.
- Sphere of Influence modeled collaborative and competitive structures between suppliers and consumers.

These simulations demonstrated characteristics of adaptive agents with emerging behaviors that are interpretable within theoretical frameworks.

Computer experiments with InfoSUMERS demonstrate the usefulness of simulation as a laboratory for computer-based experiments. The investigation of Simmel's theoretical contentions about the behavior of individuals within the fashion system show that agent-based modeling can be used to clarify and extend theoretical concepts and generate hypotheses for further research and testing. These results demonstrate the usefulness of agent-based modeling as a method of inquiry.

Virtual Consumer was conceived as a tool for product development and consumer behavior research. This individualizable "virtual consumer" allow executives to submit a proposed product to a hypothetical consumer to determine the connections between the two entities that would facilitate a purchase decision. Interest among executives of e-commerce and traditional textile/apparel businesses in Virtual Consumer shows the promise of agent-based modeling as a decision support tool for industry executives

While the efficacy of using simulations of agent-based models as an aid in strategizing about evolving markets has not been fully demonstrated, both InfoSUMERS and Sphere of Influence are suitable for use in industry and in a business school environment to develop and present cases. The reason for using simulations of agent-based modeling is that interpretation of emergent behavior leads to new strategies, tactics, and questions for empirical research.

8 Future Directions

Agents in the simulations InfoSUMERS and Sphere of Influence are relatively simple but the interactions between agents in a multi-agent simulation are complex. InfoSUMERS focuses on the information environment of consumers while Sphere of Influence focuses on the rivalry and alliances among firms in the marketplace as they target consumers with particular preference profiles. Virtual Consumer is a composite agent forming purchase intention about a particular product using a complex tradeoff analysis involving preferences, need, risk, and price/value. The next step in this research program would be to merge two simulations. Merging InfoSUMERS and Virtual Consumer would create a simulation focused on simulating cultural transmission and social dynamics with both a complex social system and agents with complex decision-making structures. Merging Sphere of Influence and Virtual Consumer would create a simulation focused on marketplace dynamics with firms competing for the attention and loyalty of agents with complex preference profiles.

The next generation of simulations—highly complex "consumer" agents in a multi-agent social environment—moves the endeavor closer to a decision support function. In the 21st Century, information from multiple sources flows continuously into the data warehouses of apparel companies. To make the huge volume of information useful in decision making, models must be in place to integrate the data, produce a holistic view of customer preferences and behavior, and display data visually. Simulation of agent-based models is particularly suitable for visualization of fashion change in social systems because agents co-evolve, adapting to each other and to changes in their world. Connecting the next generation of agent-based simulations to the data warehouse would allow apparel executives to initiate the simulation using historic data, analyze present marketplace dynamics, and explore future scenarios. Such a simulation would be capable of accurate, robust analysis and evaluation of a customer (market of one) or a market segment.

9 References

Adami, C. (1998). *Introduction to Artificial Life*. Springer Verlag.

Anderson, P.W., Arrow, K. & Pines, D. (Eds.) (1988). *The economy as an evolving complex system*. Redwood City, CA: Addison-Wesley.

Axtell, R. L. (2000). Why Agents? The varied motivations for agent computing in the social sciences. The Brookings Institution: Center on Social and Economic Dynamics Working Paper no. 17.

Axtell, R. L. & Epstein, J. M. (1996). *Growing artificial societies*. Washington, D.C.: The Brookings Institution.

Behling, D. (1992). Three and a half decades of fashion adoption research: What have we learned? *Clothing and Textiles Research Journal, 10* (2), 34-41.

Benvenuto, S. (2000). Fashion: Georg Simmel. *Journal of Artificial Societies and Social Simulation, 3*(2), http://www.soc.surrey.ac.uk/JASSS/3/2/forum/2.html.

Brannon, E.L. (1993). Affect and cognition in appearance management: A review. . In S. J. Lennon & L. D. Burns (Eds.), *Social Science Aspects of Dress: New Directions* (pp. 82-92). Monument, CO: International Textile and Apparel Association.

Brooks, R.A. & Maes, P. (Eds.) (1995). *Artificial Life IV*. Cambridge, MA: MIT Press.

Burkhart, R. (1994). The Swarm multi-agent simulation system. Position Paper for the 1994 OOPSLA Workshop on "The Object Engine."

Crane, D. (1999). Diffusion Models and Fashion: A Reassessment. *Annals of the American Academy of Political & Social Science, 566* (November), 13-25.

Cyert, R.M. (1988). *Economic theory of organization and the firm*. New York: New York University Press.

Dawkins, R. (1989). The Evolution of Evolvability. In C. G. Langton (Ed.), *Artificial Life* (pp. 201-220). Redwood City, CA: Addison-Wesley.

Day, R.H. & Sunder, S. (1996). Ideas and work of Richard M. Cyert. *Journal of Economic Behavior & Organization, 31*, 139-148.

Dewdney, A. (1988). *The armchair universe*. New York: W.H. Freeman.

Drogoul, A. (1993). *De la simulation multi-agents a la resolution collective de problems*. PhD thesis, L'Universite Paris VI.

Farrell, W. (1998). *How hits happen*. New York: Harper.

Gatignon, H. & Robertson, T.S. (1985). A propositional inventory for new diffusion research. *Journal of Consumer Research, 11*, 849-867.

Gilbert, N. & Doran, J. (Eds.) (1994). *Simulating societies: The computer simulation of social phenomena*. London: UCL Press.

Gutowitz, H. (1995). Artificial-Life simulators and their applications. Santa Fe, NM: Santa Fe Institute working paper.

Hamilton, J.A. (1988). The interplay of theory and research in discerning social meaning in dress: Implications for understanding in merchandising. In R.C. Keen (Ed.), *Theory building in apparel merchandising* (pp. 32-39). Lincoln Nebraska: College of Home Economics.

Holland, J. (1975). *Adaptation in natural and artificial systems*. Ann Arbor, MI: University of Michigan Press.

Holland, J. H. (1995). *Hidden order: How adaptation builds complexity*. Reading, MA: Helix Books (Addison-Wesley).

Holland, J.H. and Miller, J.H. (1991). Artificial adaptive agents in economic theory. *American Economic Review Papers and Proceedings, 81* (May), 365-370.

Kaiser, S. (1990). *The social psychology of clothing: Symbolic appearance in context* (2nd Ed.). New York: Macmillan.

Khalil, E. L. (1999). Two kinds of order: Thoughts on the theory of the firm. *Journal of Socio-Economics, 28* (2), 157-174.

Langton, C. G. (1986). Studying Artificial Life with cellular automata. *Physica D, 22*, 120-149.

Langton, C. G. (Ed.) (1989a). *Artificial Life*. Redwood City, CA: Addison-Wesley.

Langton, C. G. (Ed.) (1994). *Artificial Life III*. Redwood City, CA: Addison-Wesley.

Langton, C. G., Taylor, C., Farmer, J. D. & Rasmussen, S. (Eds.) (1992). *Artificial Life II*. Redwood City, CA: Addison-Wesley.

Levy, S. (1992). *Artificial Life*. New York: Pantheon Books.

Nagel, K., Barrett, C.L., & Rickert, M. (1996). Parallel traffic micro-simulation by cellular automata and application for large scale transportation modeling. Los Alamos, NM: LANL Research Paper (LA-UR-96-50).

Parker, M. T. (2001). What is ASCAPE and why should you care. *Journal of Artificial Societies and Social Simulation, 4* (1), http://www.soc.surrey.ac.uk/JASSS/4/1/5.html.

Polegato, R. & Wall, M. (1980). Information seeking by fashion opinion leaders and followers. *Home Economics Research Journal, 8*, 327-338.

Polhemus, T. (1996). *Style surfing*. London: Thames and Hudson.

Ray, T. (1992). An approach for the synthesis of life. In Langton, C. G., Taylor, C., Farmer, J. D. & Rasmussen, S. (Eds.) *Artificial Life II* (pp.371-408). Redwood City, CA: Addison-Wesley.

Resnick, M. (1994). *Turtles, termites, and traffic jams: Explorations in massively parallel microworlds*. Cambridge, MA: The MIT Press.

Reynolds, C. (1987). Flocks, herds and schools: A distributed behavioral model. *Computer Graphics, 21* (4), Proceedings of SIGGRAPH '87, 25-34.

Rogers, E.M. (1995). Diffusion of innovations. New York: Free Press.

Rogers, E.M. (1962). Diffusion of innovations. New York: Free Press of Glencoe.

Seror, A.C. (1994). Simulation of complex organizational processes: A review of methods and their epistemological foundations. In Gilbert, N. & Doran, J. (Eds.), *Simulating societies: The computer simulation of social phenomena* (pp.19-40). London: UCL Press.).

Simmel, G. (1904, October). Fashion. International Quarterly, *10* (1), 130-155.

Terna, P. (1998). Simulation tools for social scientists: Building agent based models with SWARM. *Journal of Artificial Societies and Social Simulation, 1* (2), http://www.soc.surrey.ac.uk/JASSS/1/2/4.html.

Tesfatsion, L. (1997). How economists can get Alife. In Arthur, W. B., Durlauf, S. N. & Lane, D. A. (Eds.). *The economy as an evolving complex system II* (pp. 533-564). Reading, MA: Addison-Wesley.

Waldrop, M. M. (1992). *Complexity*. New York: Simon and Schuster.

Chapter 6 Generating a Rule Set for the Fiber-to-Yarn Production Process by Means of an Efficiency-based Classifier System

Stefan Sette, Lieva Van Langenhove

Department of Textiles

University of Gent

Summary: One of the important production processes in the textile industry is the spinning process. Starting from cotton fibers, yarns are (usually) created on a rotor-spinning machine. The spinnability of a fiber is dependant on its quality and on the machine settings of the spinning machine. It would be a great benefit to be able to predict the spinnability and resulting strength of the yarn starting from a certain quality and from machine settings. To this end, two totally different modeling approaches can be considered: the so-called 'white' modeling and the so-called 'black box' modeling. In white modeling the process is described by mathematical equations, which are based upon (theoretical) physical knowledge of the process. Extensive physical information about the process is in this case available through physical, chemical or mechanical equations giving the user a thorough insight into the operation of the process. However, due to the large input (and output) dimensions of the fiber-to-yarn process and their complex interactions, no exact mathematical model of a spinning machine is known to exist nor is it likely that such a model will ever be constructed. A black box model, in contrast to white modeling, simply connects input parameters to the output without giving or containing any substantial physical information about the process itself. Black box models have been successfully constructed by Pynckels et al. to predict the spinnability (Pynckels, 1995) and the characteristics (Pynckels, 1997) of the yarn using neural networks with a Backpropagation learning rule. Apart from the lack of physical information, these models also have no fault indication or measure of uncertainty about the results.

This research will present a new modeling approach (called 'Efficiency based Classifier System' or ECS) to the fiber-to-yarn process by using an automated learning method to generate rules which allows to predict the spinnability and strength of the yarn based upon fiber quality and machine settings. This kind of modeling could be called 'gray modeling' as not only a relationship between input and output parameters is established but also physical interpretable information is generated (in this case represented as a rule set). The first part of this text will introduce the basic concepts of Efficiency based Classifier Systems (ECS). In a subsequent phase ECS will be expanded to a 'Fuzzy Efficiency based Classifier system' (FECS), which will be used to generate a rule set to predict the (continuous) yarn strength (in case of spinnability). Almost all existing applications of Learning Classifier Systems focus around a basic discrete example (e.g. the 'toy' multiplexer environment (Goldberg, 1989), etc.), which eliminates, because of its simplicity, the problem of finding a useful coding for the classifier system. Continuous environment variables can only be used when some (ingenious) mapping is done towards a binary coding. The mapping is mostly much less evident and will largely depend upon the resourcefulness of the human encoder. Using fuzzy logic to code into a number of classes (5, 7 or more) could largely solve this problem (Bonnarini, 1997). However, this requires introducing the concept of (continuous) membership degree. Each (continuous) data sample of the (fiber-to-yarn) data set, which is presented to the ECS thereby, has to be compared to all (discrete) rules of the ECS and the corresponding membership degree calculated. To this end, the matching and learning mechanism of the ECS have to be thoroughly modified into the aforementioned FECS.

The data set for the fiber-to-yarn process was selected from the database described by Sette et al. (Sette, 1997) on which also the spinnability models of Pynckels et al. (Pynckels, 1995, 1997) were based. This database was originally build within the framework of a BRITE/EURAM project (Brite, 1990) and consisted of twenty different cotton types for which five different spinning machine settings were systematically changed.

For all these settings it was experimentally determined whether the fiber was spinnable or not, resulting in a data set of 1260 spinnable and 900 unspinnable experiments. For FECS the whole fiber-to-yarn database (2160 data samples) was used with regard to the prediction of spinnability, while the 1260 spinnable experiments were used in the prediction of yarn strength.

1 Introduction Learning Classifier Systems

Unlike the 'traditional' optimization algorithms, the Genetic Algorithm is a global and robust optimization method, based on the evolutionary behavior of biological life forms in nature, as described by Charles Darwin (Darwin, 1859). A collection of points/dots in the search space is considered as a population of 'living' beings in an artificial world. The chance of survival of the fittest individual is guaranteed by elementary Darwinian guidelines. Although an element of (randomness) / arbitrariness forms part of these guidelines, the behavior is certainly not probabilistic. Historic information concerning the general purpose, 'survival of the fittest', is contained within each individual.

GAs have been developed by John Holland, his colleagues and his students at the University of Michigan in the sixties and seventies (Holland, 1973). An important breakthrough in the popularity and the use of GAs only arose after the publication of the book "Genetic Algorithms in Search, Optimization and Machine Learning" by David Goldberg in 1989 (Goldberg, 1989). Since then GAs are considered as a promising optimization technique and they are used more and more in a wide variety of optimization applications.

Next to the (obvious) use of GAs as an optimization technique to determine the global optimum of a function, this technique can also be applied to other fields, where the robustness and the global character of the optimization are important factors. A possibly interesting application for the use of GAs is the so-called 'rule based machine learning', here abbreviated as RBML. The goal is to create a collection of rules via an automatic learning process that allow a system ('machine') to function optimally in a particular environment. This can be schematically represented as in Figure 1.

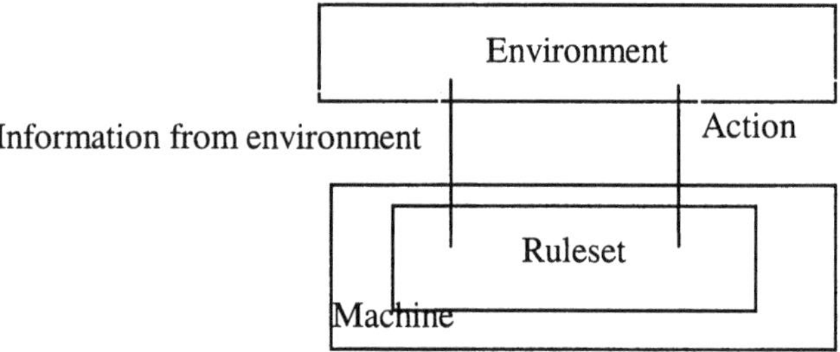

Figure 1: Representation of RBML configuration

Each rule has the form {IF condition THEN action}, in which the condition as well as the action can differ in length, form and contents. The final collection of rules searched for must have the following properties:

- implementing an optimal strategy for the machine with regard to its environment
- not containing any superfluous information and consequently forming a minimal set of rules.

So, the construction of a set of rules for RBML comprises in fact a twofold optimization problem. Depending on the complexity of machine and environment, both optimization problems can be subject to a great number in the search space, local optima, etc., and consequently form a typical problem for a GA.

The application of a GA in the search for an optimal set of rules in machine learning is known under the name 'genetic based machine learning', abbreviated here as GBML. The originally most common GBML architecture is the so-called 'learning classifier system', abbreviation LCS, developed by Holland [Holland, 1978].

1.1 Learning Classifier Systems

1.1.1 Introduction

In a learning classifier system (LCS) the set of rules must have a suitable form. It must also possibly have an assessment value (with regard to the optimal functionality of the machine environment) for a GA to be applicable. An LCS will consequently be composed of the following three components:

- a system of rules and messages
- evaluation of the set of rules via an assessment mechanism (e.g. via a rewarding strategy),
- a GA to let evolve ('learning') into a 'stronger' set.

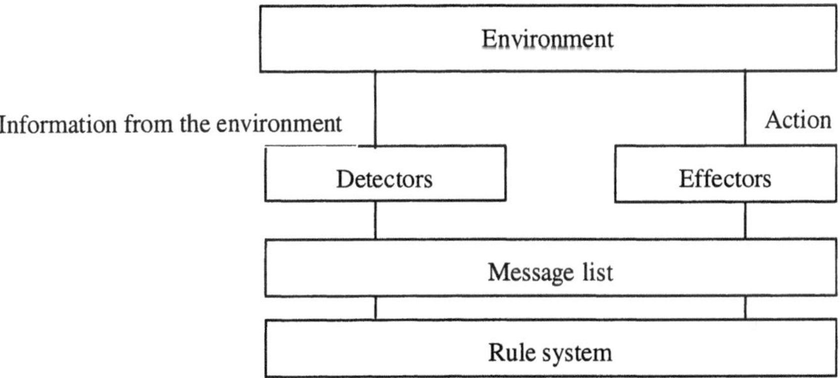

Figure 2: System of rules and messages

1.1.2 The System of Rules and Messages

The system of rules and messages forms the interaction of the LCS with its environment. Using detectors the messages are read in the list of messages, which are interpreted by means of a system of rules and finally generate a message (for the environment) via the effector. The environment message is mostly an action that affects the environment. A schematic representation of the system of rules and messages is represented in Figure 2.

A message, which has been read in, can immediately generate an (environment) action by means of a rule, which is then again placed on the list of messages. In this way different internal messages can be generated, before a rule gives rise to an environment action.

As has already been mentioned, a rule of the form:

IF<condition>THEN<action>

in which the condition (and the action) can consist of multiple components. All (coded) rules in the system of rules need to have the same length, though. This will allow generating new rules from the existing set of rules using a GA. If internal messages occur, then the length of the condition must also equal the length of the action, so that the action can be interpreted as an (internal) message on the list of messages.

If the messages and the rules (also called 'classifiers') are coded binary, the following definitions apply:

<classifier> = <condition>:<message>

<condition> = $\{0,1,\#\}^k$

<message> = $\{0,1\}^k$

where $\{0,1,\#\}^k$ or $\{0,1,\#\}^k$ stands for addition (concatenation) of the respective symbols 0, 1 and 0, 1, # to a string with a length k. Here # is the 'don't care' symbol, in other words a condition is met for # if 0 as well as 1 is present in the position concerned. Only the condition of the classifier can contain #s.

Suppose e.g.; the following classifier system ($k = 4$):

0 # # 0 : 1 1 1 1 (1)

1 # 0 1 : 0 1 1 0 (2)

1 0 : 0 1 0 0 (3)

1 0 # : 0 0 0 0 (4)

and a message {1 0 1 0}.

This message (only) matches the condition of the third classifier generating the internal message {0 1 0 0}. As the condition of classifier (1) as well as that of classifier (4) match this, the following two messages are generated: {1 1 1 1} and {0 0 0 0}. The message {0 0 0 0} leads also to message {1 1 1 1} via classifier (1). The resulting message {1 1 1 1} could be e.g. an action regarding the environment (it is silently/implicitly accepted here that a 1 on the least significant location of a message is an occasion/(immediate) cause to send this to this environment).

1.1.3 The Rewarding Mechanism

To be able to distinguish well performing classifiers from less suitable classifiers a mechanism should be implemented to assess classifiers. For this each classifier *I* at a time instant *t* is provided with a particular parameter $S_i(t)$, strength, indicating how well the classifier performs.

The adaptation of $S_i(t)$ can occur [Holland, 1978] on the basis of the so-called '*Bucket Brigade Algorithm*' (BBA). The BBA's construction is analogous with a simple economical system where one or more suitable participants with a high bid get the chance to make a 'profit' (which can also end in a loss for the participant concerned). In an LCS the participants are classifiers and their capital is the parameter $S_i(t)$. In other words: when a particular message is presented in the list of messages, the classifier with a matching condition can make a bid (e.g. proportional to their strength). The classifier making the highest bid (possibly multiple classifiers) is then activated, on condition of paying the bid. There are two possible consequences:

1. The selected classifier generates an internal message for the list of messages of the LCS. A possible reward returns in the following cycle form the bid of other classifiers on the basis of the selection of the new message.
2. The selected classifier generates an action either immediately causing from the environment a reward or not.

It should be noted that the bid made goes to the environment if the message was derived directly from the environment, or - in the other case - to one or more classifiers having generated the message.

The first possibility can lead to a chain of co-operating classifiers, a reward being finally received (in the favorable case) of the environment (second possibility). The strength of the last classifier in this chain will increase in the case of a reward. When going through this sequence again will, in this way, the final bid also be higher, also increasing the strength of the remaining classifiers in the chain. It is also clear that the BBA will not strengthen a single classifier, but a subset of classifiers that can co-operate.

The modification of the strength of a classifier at the time instant $t + 1$ by means of the BBA can be expressed as follows:

$$S_i(t+1) = S_i(t) - P_i(t) - T_i(t) + R_i(t) \tag{5}$$

with

$S_i(t)$ the strength of the classifier i at time t

$P_i(t)$ bid of the selected classifier i at time t

$T_i(t)$ tax of the classifier i at time t

$R_i(t)$ the reward of the selected classifier i at time t

with

$$P_i(t) = \alpha . S_i(t) \tag{6}$$

$$T_i(t) = \beta . S_i(t) \tag{7}$$

The bid *Pi(t)* as well as the tax *Ti(t)* are usually chosen proportional to the present strength *Si(t)* of the classifier where mostly $0 < \beta < \alpha < 1$. The tax term *Ti(t)* is necessary to reduce the strength of the classifiers that are never activated (in other words superfluous or senseless classifiers) and can be applied on all classifiers (independent of a possible bid or reward) each cycle.

Starting from an example of Section 1.1.2, formula 5, 6 and 1.3, where

$\alpha = 0.1$,

$\beta = 0.0$ (no tax component),

a environment reward for the action of 20

$\forall i, S_i(0) = 100$

Table 1 is obtained for the first 3 time steps.

Table 1: Strength development of classifiers by means of the BBA

i	$S_i(0)$	$P_i(0)$	$S_i(1)$	$P_i(1)$	$R_i(1)$	$S_i(2)$	$P_i(2)$	$R_i(2)$	$S_i(3)$
(1)	100		100	10	20	110	11	20	119
(2)	100		100			100			100
(3)	100	10	90		10+10	110			110
(4)	100		100	10		90		11	101

Classifier (3) is activated by the (environment) message {1 0 1 0}, the bid of 10 going to the environment. The internal message {0 1 0 0} then activates the classifiers (1) and (4) each bringing out a bet of 10, going to the classifier that generated the internal message in the previous cycle, this is classifier (3), now with a total strength of 110. The message {1 1 1 1}, generated by classifier (1) is an action that is rewarded immediately by the environment by 20, turning the strength of (1) into 110 as well. The internal message of (4) {0 0 0 0} activates in the following cycle again classifier (1), which gets a bid of 11 and a direct reward of 20. The bid goes to classifier (4). Finally, classifier (1) ends with the highest strength (after having sent a successful action to the environment twice). Classifier (3), which gave the original incentive to the successful action, has the second highest strength. Note that the classifier (2) never came into consideration and, consequently, still has the strength 100. Taking into account a tax-component, the strength of classifier (2) would have decreased in the meantime.

Goldberg [Goldberg, 1989] shows that the BBA is stable (does not diverge) for ending limited rewards $0 < \alpha + \beta \leq 2$ (in practice $0 < \alpha + \beta \leq 1$). The 'steady state' values (for constant rewards) of the strength and the bid are given by:

$$S_{ss} = \frac{R_{ss}}{\alpha + \beta}$$

$$P_{ss} = \frac{\alpha}{\alpha + \beta} R_{ss}$$

The mentioned implementation of the BBA (as well as in the example) started from a deterministic vision: the highest bid was always chosen to activate the corresponding classifier. This can lead to non-optimal choices in which a suitable classifier, with minor strength is always systematically ignored. Two possible variants avoiding this problem are:

the application of a noise factor on the bid (or the strength) of each classifier, so that the strength is determined by:

$$P'_i = P_i + N(\sigma_{bod})$$

where N is a noise function with standard deviation σ_{bid} and the application of a probabilistic selection of the classifier, based on its bid analogous to the roulette wheel algorithm for GA. In this way *the chance* of activation proportional to the bid of the classifier.

1.2 The GA

The system of messages and rules, as well as the assessment of the classifiers were discussed in the previous two Sections. The last stages consist of by means of the information obtained (classifier strength), create new classifiers. Thanks to the imposed classifier format and the introduction of a parameter assessing the performance of each classifier (namely the strength) the application of a GA is obvious. The classifier list now is the population (which the GA works on) the fitness of each string (classifier) being determined by the strength of the corresponding classifier. The complete LCS can be schematically represented in Figure 3.

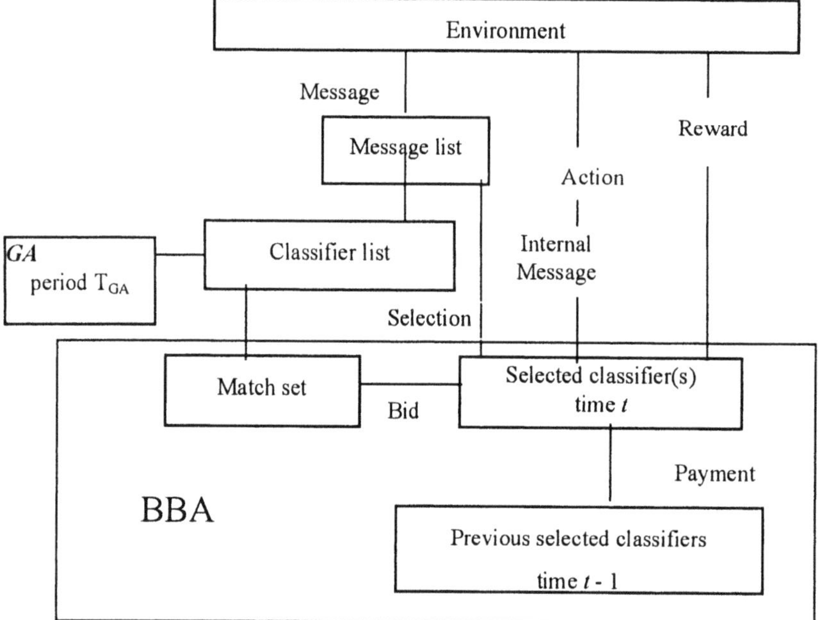

Figure 3: Overview of an LCS

It is necessary, though - before allowing the GA to generate a new classifier population - to assess the present population sufficiently (by means of the BBA), so that the strength of each classifier approaches its '*steady state*' strength. For this a number of message and rule cycles should already have been run through. The period T_{GA} for the application of the GA on the classifier population can be implemented according to three guidelines:

- deterministic, after going through a fixed number of cycles,
- probabilistically, with an average period T_{GA},
- when reaching particular performance and convergence characteristics of the present classifier set.

The construction of a GA for an LCS is essentially similar to a basic GA as discussed by Goldberg (Goldberg, 1989). Because of the special nature of the optimization in the LCS the following adaptations to should take place:

The complete replacement of the old population by a new, GA-generated population is to be avoided. In contrast to the (univocal) function optimization to one global maximum, it is the LCS's goal to generate an optimal *set* of (different) classifiers. Always replacing a full population would result in the loss of already optimal classifiers of the old population, while new (though possibly different) optimal classifiers are generated. Obtaining a full optimal classifier set can be considerably hindered by this. Consequently, it is necessary to work more selectively in the replacement of a population. An elite reproduction and crossing scheme is therefore recommended: the better strings (classifiers) of the older population must be kept.

The mutation operator will now work into a ternary alphabet {0, 1, #} for the condition of the classifier, instead of the binary alphabet {0, 1}. In other words: '0' can be mutated to a '1' or '#' (with equal chance), a '1' mutates to '0' and '#' and a '#' to '0' or '1'.

2 Efficiency Based Classifier Systems (ECS)

It can be shown (Sette, 1998) that the BBA has some fundamental problems to achieve a final (successful) classifier set in a non-ideal environment:

the strength for t going up to ∞ of a (non 100% correct) classifier knows no fixed value and can vary widely.

The choice for a constant reward R will greatly influence the classifier behavior during the first cycle of the BBA.

A possible solution for these problems can be found in:

> *An (additional) classifier parameter, necessary to assess the efficiency of each classifier. This parameter should contain information concerning performances already achieved by the classifier without being liable to important changes at the end of the BBA cycle.*

Let the reward R vary in function of a classifier population parameter which gives a guideline for the (average) success of the present population. On the basis of this parameter the instant R can then be determined in such a way that the (generalizing) classifiers with a small level of success (relative to the average success achieved in the population) are no longer competitive with this population.

In the following Sections the basic structure (2.1), the definition of efficiency (2.2), a detailed structure description (2.3) the modified BBA (2.4) and the GA implementation for ECS (2.5) are described, taking into account the guidelines described above [Boullart, 1998].

2.1 Construction of an ECS Basic Design

Figure 4 shows the basic structure of the classifier system on which ECS is based. The environment generates a message that is presented to the classifier set. When the condition part of the classifier matches with the message, the classifier is selected for the match set. This is called a *virtual selection* (of the classifier), as the classifier is now a potential candidate for activation. Using the classifier strength (via a bid in which the highest strength wins) now (only) one classifier of the match set is selected for activation. This is called a *real selection* as the classifier pays the bid and receives a possible reward, in this way changing its strength.

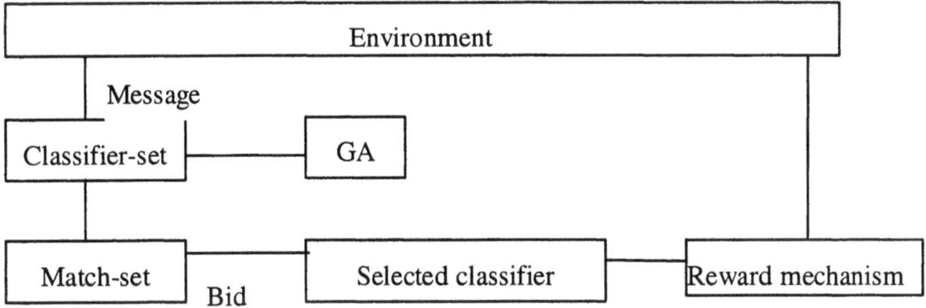

Figure 4: Basic structure of the ECS

On the basis of the environment a reward is allocated or not to the activated classifier (presupposing a one-step environment). The classifier is considered *successful* when it has received a reward and *unsuccessful* (failing) if not.

For each classifier in the classifier set 4 parameters are defined (counters, in fact) which are continually adapted during the development/course (in each cycle) of the BBA. These parameters determine respectively:

number of *real selections,*

number of *real successful* rewards,

number of *virtual selections,*

number of *virtual successful* rewards.

The last parameter (virtual successful rewards) is adapted when the corresponding classifier belongs to the match set: without activating the classifier (so independent/irrespective of its strength) it is determined whether the classifier will be considered for a (virtual) reward. If this is the case, the parameter is incremented (all other parameters, including strength, retain their original value).

Moreover, two other global classifier system parameters (counters) are defined:

number of presented messages with regard to the global classifier *system*

the number of successful rewards that the classifier *system* has received as one whole

For each selection or successful reward the corresponding parameter (counter) is incremented with one. At the start and after each GA cycle all parameters mentioned above are initialized on 0.

On the basis of these parameters the notions classifier efficiency can be defined for each BBA cycle.

2.2 Definition of Classifier Efficiencies

The efficiency of a classifier (or, more generally, the total classifier system) can be described using the following variables:

a(t): the *accuracy* of the number of accumulated successful rewards of the classifier of the classifier system.

g(t): the *generality* of the total number of accumulated selections (whether successful or not) of the classifier or the classifier system.

Efficiency *E(t)* at the time *t* is defined as:

$$E(t) = \frac{a(t)}{g(t)} \qquad \text{met} \qquad 0 \leq E(t) \leq 1.0$$

E(t) turns maximally into 1.0 when all selections (up to the time *t*)are successful and is minimally 0.0 when no single selection (up to time *t*) is successful.

The following specific efficiency parameters can be defined:

Global efficiency $E_g(t)$: the efficiency of the complete classifier system at time *t* is calculated as the ratio of all successful rewards relative to al presented environment messages. This parameter measures the efficiency of the *full* classifier system at time *t*. an efficiency $E_g(t)$ indicates a classifier system where all environment messages are responded to successfully.

Virtual efficiency $E_v(i,t)$: all classifiers *i* in the match set are virtually selected and for each of these classifiers it is determined whether they would be considered for a reward, resulting in a virtual efficiency at time instant *T*. As has been mentioned earlier these classifiers need not be effectively activated in the BBA.

Real efficiency $E_r(i,t)$: only the classifiers i that are activated by the BBA (with possible real reward) are considered when the real efficiency s calculated at time t.

2.3 Structure of the ECS

A schematic overview of the operation of the ECS is given in Figure 5.

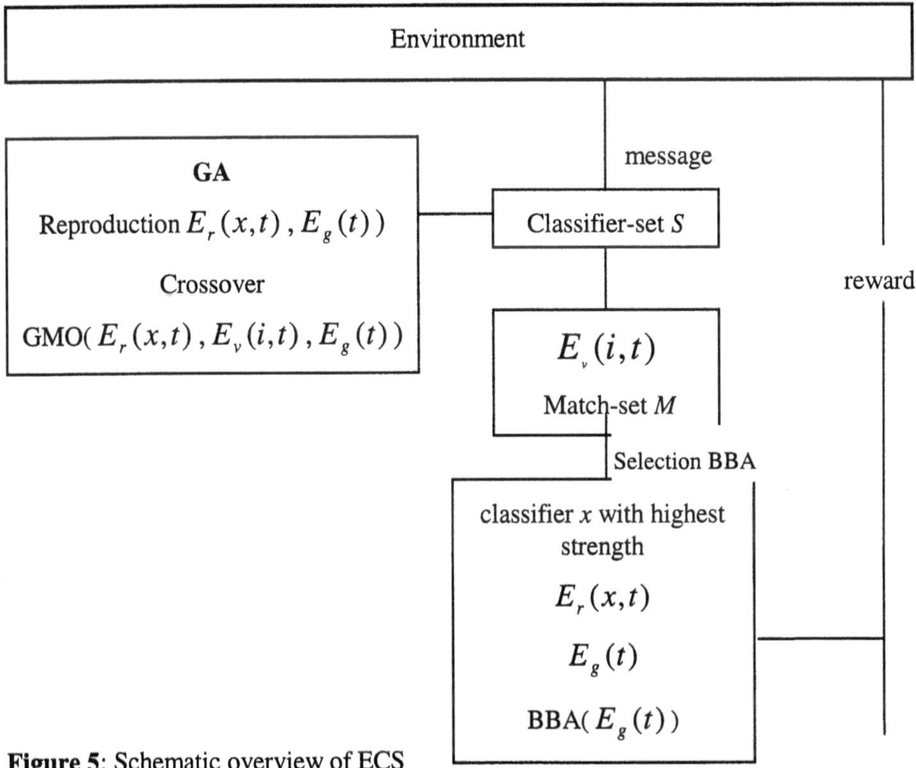

Figure 5: Schematic overview of ECS

On the basis of the definitions in Section 2.2 and Figure 5 an elementary ECS cycle can now be described. This runs as follows:

The environment generates a message represented to the classifier set S of the classifier system.

From the classifier set S the corresponding (condition part equals message) classifiers are selected that form the match s M. During this step the virtual efficiencies $E_v(i,t)$ of the classifiers form M are determined.

From the match set M the classifier x with maximum strength is activated by the BBA and checked on possible reward by the environment. At this moment the real efficiency $E_r(x,t)$ (for the activated classifier) as well as the global efficiency $E_g(t)$ (for the complete classifier system) are adapted. The possible reward R is calculated on the basis of $E_g(t)$ (see Section 2.4: the efficiency modified BBA).

After a fixed number of BBA cycles the GA is activated (working into the classifiers belonging to S) to introduce in this way new classifiers in S. To do this the GA bases itself on the classifier strength, $E_r(x,t)$, $E_v(i,t)$ and $E_g(t)$ (see Section 2.5: GA implementation for ECS).

If ECS is compared to the original algorithm of Holland (see Section 1.1), then the following differences become clear:

- for ECS, the reward of the BBA is based on the global efficiency of the classifier system,
- for ECS, the GA is a function of all three defined efficiencies.

In the following two Sections these efficiency based algorithms are more elaborately discussed.

2.4 The Efficiency Modified BBA

To avoid over (generalized) and low efficiency classifiers it is desirable to presuppose the reward R variable and dependent on the efficiency of the classifier population. An approach for the average efficiency of the (active) classifier population at time t is given by the global efficiency.

This is adapted as already mentioned earlier in Section 2.3 in each BBA cycle so that already from the first activated classifier (first BBA cycle). During the further development of the BBA an advanced refinement will take place offering a more accurate approach for the average efficiency of the (active) classifier population. At the start of the BBA the strengths of the classifier population are initialized at a fixed value. Note that this occurs several times during the learning process of the ECS, namely after every application of the GA on the classifier population.

The reward in the modified BBA is consequently defined as:

$$R(t) = \sigma - \frac{E_g(t)}{\varepsilon} \tag{8}$$

met $\sigma = 1.0$ en $\varepsilon = 3.0$.

2.5 GA Implementation for ECS

After running through a fixed number of BBA cycles, the GA will generate a new population, based on the fitness of the classifiers from the previous generation of classifiers. The following Sections discuss respectively the ECS implementation for the reproduction, crossing and mutation operator.

2.5.1 Reproduction ($E_r(x,t)$, $E_g(t)$)

In a (random) classifier system replacing 100% of the old classifier population by a new classifier population is to be avoided. A complete elimination of the old classifier population would also eliminate the already generated, correct (and optimal) classifiers. Consequently, it is necessary to keep a number of successful classifiers in a new population originating from the old classifier population. This limited collection of successful classifiers needs to be dynamic and the number of classifiers in this collection must be determined by the strength and efficiency of the classifier. A possible, simple algorithm could be based on a particular limiting value for the strength (or the efficiency): each classifier of which the strength or efficiency would score higher than the fixed limit would be taken up into the new population. This solution is too static, though, as the strength (and efficiency) could change drastically during the learning process of the classifier system. This can result in hardly any (or no) classifiers with a high strength (and efficiency) at the beginning of the learning process and a great number of classifiers with a high strength (and efficiency) at the end of the learning process.

Classifiers with a high strength and high efficiency, who are to be inserted into the new population, must be selected on basis of the global efficiency E_g of the current classifier population. This procedure insures the existence of classifiers with a (relative) high strength and a (relative) high efficiency during the whole learning process in each classifier population.

The following elitist algorithm is used to determine the classifiers with high strength and efficiency within each population:

the classifiers are ordered according to strength, starting with the highest strength, the real efficiency $E_r(i,t)$ of the corresponding classifier i is compared to the global efficiency $E_g(t)$ of the population:

If ($E_r(i,t) > \mu.E_g(t)$) with $0 < \mu < 1$) then the classifier i is selected to be copied into the next generation (9)

Again this algorithm combines the classifier strength (calculated using the BBA) with the concept of classifier efficiency (accuracy or generality of the classifier). In the ideal case, at the end of the learning phase this would lead to the selection of the minimal optimal classifier set.

The reproduction operator is, with exception of the aforementioned elitist selection of classifiers with a high strength and high efficiency, implemented as described by Goldberg (Goldberg, 1989) using the roulette wheel algorithm. A guideline for the composition of the full classifier population (elitist reproduction versus standard reproduction and crossover) is given in Section 2.5.3

2.5.2 Crossing

A standard implementation independent, of the efficiencies was used.

2.5.3 Guided Mutation Operator (GMO)

The GMO is a new mutation operator that can replace or complement the existing (GA-) mutation operator (as described in Section 2.3).The GMO is based on the *(virtual) accuracy* and the *(virtual) generality* of each classifier. The purpose of the classifier system can be described as the combination of a maximal accuracy combined with a maximal generality using a classifier set as small as possible. This concept allows constructing 'probably' better classifiers based on the previous population by using a new mutation operator.

When the $a_v(i,t)$ is the virtual accuracy of classifier i at the time instant t and $a_g(i,t)$ the virtual generality of the classifier i at the time t , then the following possibilities can be distinguished:

- $a_v(i,t) < a_g(i,t)$: the corresponding classifier i at the time t has no maximal accuracy and should be adapted in order to improve its accuracy.
- $a_v(i,t) = a_g(i,t)$: the corresponding classifier i at the time t has a maximal accuracy, but can be adapted in an attempt to increase its generality.

A mutation operator (GMO) can now be developed, going through the following steps:

- **if** $a_v(i,t) < a_g(i,t)$: **then** one # ('don't care') symbol of the classifier i is selected at random and replaced by 0 or 1. This reduces the generality of the classifier and 'hopefully' improves the accuracy.
- **If** $a_v(i,t) = a_g(i,t)$ **then** one specific classifier location (0 or 1) of the classifier i is selected at random and replaced by a # symbol. This increases the generality of the classifier.

The GMO should preferably be applied to successful classifiers (with a high strength and high efficiency) to introduce (mutated) classifiers with an even higher accuracy or generality. The selection of the classifier-'parents' for mutation is therefore similar to the selection procedure for the (elitist) reproduction of classifiers with a high strength and high efficiency (see Section 2.5.1). In other words, the (mutation) selection happens on the basis of ordered strength and real efficiency.

Supposing that n classifiers with a high strength and high efficiency are suitable/eligible for elitist reproduction in the new classifier population, then the same n (successful) classifiers will form the basis for n new (mutated) classifiers. It is clear that within the classifier population sufficient space is needed for the classifiers generated by normal reproduction and crossing. The optimization algorithm would be considerably limited if the new classifier population only consisted of classifiers with high strength and high efficiency and the mutations following from this. In the starting phase of the optimization algorithm this could lead to a small (possibly locally optimal) group of classifiers that, dominate the entire population, together with their mutations. This could lead to finding the local optimum instead of a global optimum as the population variety is considerably limited. In practice this means that the total number of classifiers in the population should always be higher than $2n$ and preferably twice the size so that the elitist classifiers and their mutations only represent half of the population. The full classifier population in balance can be composed as follows:

- 25% of the classifier of the previous (old) population undergoes elitist reproduction in the new population
- 25% of the classifiers in the new population consist of a mutation of the previous (elitist) group of classifiers
- 50% of the classifiers in the new population are generated via the classic reproduction using the roulette wheel algorithm and the classic crossing operator.

When this balance cannot be reached (in other words, the two first contributions comprise more than 50% of the total population), a larger classifier population should be selected during initialization.

3 ECS-implementation for the Generation of Industrial Production Rules

3.1 The Fiber-yarn Production Process

As described by Sette et al. (Sette, 1998), the modeling of the fiber-yarn process using neural networks is a '*black box*' model linking only input parameters to output parameters without generating substantial information about the process itself. Moreover, there is no error indication or measure of (un)certainty about the attained results (apart from the relative percentile error). This application aims at generating rules for the fiber-yarn process to predict spinnability based on the fiber properties and machine settings.

The following subset of the fiber-yarn data presented (Sette, 1998) was therefore taken as a basis:

- 4 different fiber qualities selected on the basis of different fiber strength.
- 5 machine settings (yarn count, twist, navel, breaker and rotor speed) with different discrete settings.

The output consisted of the spinnability (yes/no) of the subset mentioned above. The complete learning file consisted of 426 different series of fiber-yarn data.

3.2 Construction of the ECS

The classifier-format is represented in Table 2.

Table 2: Classifier-format for the fiber-yarn process

	fiber strength	**yarn count**	**Twist**	**navel**	**breaker**	**Rotor**	**output**
Settings	4	3	3	2	3	2	2
Coding	2 bits	2 bits	2 bits	1 bit	2 bits	1 bit	1 bit

The first row gives the input parameters (1 fiber property and 5 machine settings) and the output parameter (the spinnability of the fiber and machine settings). The various matching settings and the number of bits necessary for the (binary) coding are given in the second and third row respectively. Consequently, the condition part of the classifier is coded in 10 bits and the action part in 1 bit.

Formulae 5, 8 and 2.2 were used with

α=2/3, β=0.01, σ=1.0, ε=3.0 and μ=0.5

During initialization the classifier content is determined at random and a strength is (at random) attributed between 0.9 and 1.1. For each new population (after application of the GA-process) the strength was again randomly initialized with the exception of the classifiers with high strength and high efficiency (which were reproduced in an elitist way in the new population) receiving a strength bonus of 0.2 on top of the random initialization (resulting in a strength between 1.1 and 1.3). Generality and accuracy were again initialized at 0 for all classifiers.

The ECS experiment for the fiber-yarn process was executed on the basis of a classifier population of 100 classifiers and 150,000 cycles. The learning file was run through three times using the modified BBA algorithm (1278 BBA cycles) after which the GA generated a new classifier population. The number of GA cycles amounted to 117.

After the learning phase, the final classifier selection was executed on the classifier population (of 100 classifiers), in which the corresponding classifier i was selected for each message meeting the following conditions:

the highest strength

AND ($E_v(i, t) > \mu . E_g(t)$ for $t = 150{,}000$ and μ determined by the user (with $0 < \mu \leq 1$)

maximal efficiency (100%) can be reached by basing the selection on $E_v(i, t)= 1$ instead of $E_v(i, t) > \mu . E_g(t)$. The corresponding classifiers will never fail in activation (real as well as virtual). In other words, the resulting classifier set will be entirely faultless with respect to the presented environment messages.

The selection criterion $E_v(i, t) > \mu . E_g(t)$ is less selective: depending on the value μ a choice can be made to reach a higher or lower classifier efficiency. In the fiber-yarn experiment the following values are considered: μ = 0.9 and μ =0.8. Generally, it can be presupposed that at the end of a (successful) learning process the global efficiency $E_g(t)$ will be close to 100%, so that the finally selected classifier set will have a virtual efficiency of about 90% respectively 80%.

3.3 Results of the ECS for the Fiber-yarn Process

The results are summarized in Table 3. The first column shows the classifiers with a maximum virtual efficiency (=1.0), while the second and third column match the decreasing virtual efficiency (μ = 0.9 and μ =0.8). The first row (number of classifiers) gives the number of selected classifiers according to the criterion given in the columns. The total % correct predictions gives the total number of experiments (messages) in which a correct prediction (action) was taken by the corresponding classifier set (this considering the 426 experiments). The efficiency shows if the corresponding classifier set has made errors, while the prediction area indicates the (%) number of experiments to which the classifier can be applied. Finally, the number of predicted experiments is classified according to their spinnability combined with the prediction by the classifier set -whether this is correct or not.

Table 3: Spinnability in the fiber-yarn process

	$E_v(i, t) = 1$	$\mu = 0.9$	$\mu = 0.8$
number of classifiers	26	32	36
total % correct predictions	89%	92.7%	94.1%
efficiency (E_v)	100%	95.2%	94.1%
prediction area	89%	97.4%	100%
total number of experiments	426	426	426
non-spinnable(correct)	102	106	110
non-spinnable (incorrect)(1)	0	3	5
spinnable (incorrect)(2)	0	17	20
spinnable (correct)	277	289	291

(1) these are the non-spinnable yarns which were predicted to be spinnable
(2) these are the spinnable yarns predicted to be non-spinnable

The following conclusions can be drawn:

The highest number % correct predictions is generated by the greatest classifier set (μ = 0.8), but this corresponds also with the lowest efficiency. 94.1% of all experiments was predicted correctly. This can be divided into 95.7% accuracy for the non-spinnable fibers and 93.6% accuracy for the spinnable fibers. These results are comparable to and even exceed the attained accuracy of back propagation NN-model [Pynckels, 1995], in which an accuracy 95% was reached for the unspinnable fibers and 90% accuracy for the spinnable fibers.

The highest efficiency (100%) is attained by the smallest classifier set, but has a prediction area of only 89% regarding the total number of experiments. In other words, the total number of predictions becomes smaller than that of the corresponding back-propagation NN-model, though the predictions always have 100% efficiency (completely free of errors considering all examples presented to the ECS). This is a direct consequence of the selection condition $E_v(i, t)=1$.

Table 4: 'Infallible' classifiers for the fiber yarn production process

Fiber strength		Yarn count		Twist		Navel	Breaker		Rotor	Output	Strength	Selection	Reward
1	0	0	#	#	#	1	#	#	0	1	0.668	2	2
#	0	0	0	0	#	0	1	#	0	0	0.062	2	2
0	#	0	0	0	0	0	0	#	1	1	0.501	2	2
1	0	0	#	0	#	1	#	#	1	0	0.187	2	2
0	#	#	#	1	#	#	#	0	0	1	0.563	2	2
0	#	0	#	#	#	#	0	0	0	1	0.051	3	3
#	0	0	0	0	#	1	#	#	0	1	0.961	4	4
#	#	#	#	1	#	0	#	1	#	1	0.362	4	4
0	#	0	1	#	#	0	#	#	#	1	0.749	6	6
1	0	1	#	#	#	#	1	#	#	1	0.882	6	6
0	1	#	#	#	1	0	#	#	#	1	0.828	6	6
1	#	1	0	#	1	0	0	#	#	1	0.811	8	8
1	1	#	0	0	0	0	#	#	#	0	0.859	9	9
1	#	0	0	0	#	0	#	#	0	0	1.025	12	12
1	#	0	0	0	#	1	#	#	1	0	1.006	12	12
#	0	0	0	0	#	0	#	#	1	1	0.957	12	12
#	#	0	1	#	#	1	0	#	1	0	0.873	12	12
#	#	0	1	#	1	0	#	#	#	1	0.845	12	12
#	1	#	#	1	#	0	#	#	#	1	0.611	12	12
0	1	#	#	#	#	1	#	#	0	1	0.998	14	14
#	#	0	#	1	#	#	#	#	0	1	0.899	30	30
0	#	1	#	#	#	0	#	#	#	1	0.992	36	36
1	#	1	#	#	#	1	#	#	#	1	0.991	36	36
#	#	#	1	#	#	0	#	#	1	1	1.01	36	36
#	#	0	#	#	#	1	0	#	0	1	1.01	46	46
0	#	#	#	#	#	1	#	#	1	0	0.93	53	53

The faulty predictions (for $\mu = 0.9$ and $\mu = 0.8$) are made especially in spinnable experiments that are (incorrectly) classified as unspinnable. These mistakes are easier to accept in the real production process than the reverse: classifying non-spinnable fibers as spinnable. The latter would lead to starting the production process without being able to spin a yarn.

The most interesting result is generated using the classifier set determined by $E_v(i, t)$= 1, resulting in 26 'infallible' rules concerning the presented dataset. Table 4 gives an overview of this set of infallible classifiers.

These 26 classifiers describe 89% (for a total of 379 selections) of all experiments with an efficiency of 100% (selection = reward). Analyzing Table 4 further, the following conclusions can be drawn:

55.6% (or 62.5% if only the prediction area of the classifier set is considered) of all experiments is predicted by (only) 6 classifiers.

Each classifier has a real significance. E.g.: consider the classifier with the highest number of selections represented by the code:

{0# ## ## 1 ## 1 : 0}

The interpretation of this is as follows (see the classifier-format in Section 3.2, Table 2):

> "if the fiber strength is very small (00) or small (01) and a carved navel (1) is used, as well as a high rotor speed (1) then the yarn count (##), twist level (##) and breaker speed (##) are unimportant, the spinning machine will not be able to spin the yarn (0)."

This offers important information on the operation of the fiber-yarn production process and the way in which this process should be configured.

The input parameters can be classified further according to their importance in the set of rules mentioned above by summing all selections with a 'significant' input parameter (consisting of at least one 0 or 1) in a classifier. The result of this is represented in Table 5.

Table 5: Importance of input parameters

	Fiber strength	Yarn count	Twist	Navel	Breaker	Rotor
Significant selections	239	288	129	338	85	244
order of importance	4	2	5	1	6	3

This shows that navel is the most important parameter in the generated classifier set with 338 significant selections, followed by yarn count, rotor speed, fiber strength, twist and breaking speed.

3.4 First Conclusions Concerning ECS

The original BBA (as applied in the LCS of Holland) appeared to be inadequate as measure for the final selection of a classifier when a non-ideal environment is to be modeled. Moreover, it is clear that a constant reward will influence the classifier behavior strongly (in particular the selection of certain classifier efficiencies) from the first cycles of the BBA.

To face the problems mentioned above, the concept (classifier-) efficiency was input to the ECS model on three different levels: real, virtual and global efficiency. The GA and the BBA of the ECS-model were changed, using the different efficiencies. In the GA an elitist reproduction was introduced on the basis of real and global efficiency, as well as a guided mutation operator that is a function of the accuracy and generality of the classifier to be mutated. In the BBA the reward was made dependent on the global efficiency.

The ECS configuration was confronted with a non-ideal environment in the shape of the fiber yarn production process. Numerous classifier sets were generated in this way, their prediction error always being smaller than 6%. Besides, physical information was gathered in the shape of interpretable rules on the observed production process. These rules, deduced directly from the generated classifier set, give information concerning the fiber strength and machine settings to be selected to achieve a spinnable yarn.

The following chapter will alter the ECS-model so that also continuous input and output parameters (of an industrial process) can be processed. This will lead to generating sets consisting of fuzzy rules.

4 Development/Extension of ECS for Continuous Parameters

This chapter studies the problems in processing continuous parameters by ECS (and LCS). The introduction of fuzzy classes (and corresponding degrees of belonging/attachment) changes ECS to the 'efficiency based fuzzy classifier system' (FECS). The achieved model is again verified using the fiber yarn process (in particular for the prediction of the spinnability). Finally the number of fuzzy classes necessary for a continuous prediction in the fiber yarn process is studied (in particular the prediction of the yarn strength).

4.1 Introduction

ECS (and other LCS configurations) are suitable for the generation and evaluation of rules that are composed from discrete parameters. A simple example of this is the multiplexer problem where all information (rule coding for input and output parameters) is presented in a binary way. As the problem itself is completely binary no conversion is consequently needed to achieve the coding of the classifiers. This is also the case for the fiber yarn production process as was studied in chapter 5, Section 5.4. Each machine parameter only has 2 or 3 possible discrete settings, so that a binary representation is obvious. But the great majority of the 'real' environments and processes has continuous parameters where the conversion to a functional classifier format is less evident. In the following Sections a solution is suggested in which the continuous parameters are divided in '*fuzzy sets*' and the ECS algorithm is extended for the processing of classifiers on the basis of degrees of belonging.

4.2 Fuzzy Collections Applied to ECS Problems

Schematically an LCS can be presented as in Figure 6.

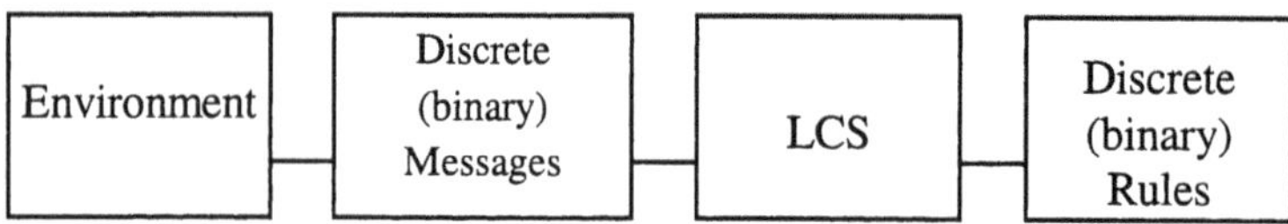

Figure 6: Input and output for an LCS

Many LCS applications will concentrate on 'discrete' problems (such as the multiplexer environment) where a suitable coding of the classifiers is obvious. Continuous environment functions can (only) be implemented when a (often not evident) transformation occurs to a binary coding (where the problems mentioned in the previous Section must be avoided). Often the transformation will depend on the inventiveness of the human coder.

The implementation of fuzzy sets where the parameter is split in a number of fuzzy classes (5, 7 or more), can offer a solution for many applications. The concept of continuous degrees of belonging are also introduced here:

Each (continuous) message of the environment that is presented to the ECS can be compared to all (discrete) classifiers (belonging to the classifier list) where a corresponding membership degree is calculated. The membership degree is a measure for the applicability of the classifiers towards the presented example.

To realize this the matching -and learning mechanism of the ECS must be adapted.

The description of FECS is here explained with the use of two classes. In Section 4.3.2 the FECS is then further extended (and studied) for numerous classes.

Suppose an environment message

$$M(p_1^I, p_2^I, \dots, p_n^I, p_1^O, p_2^O, \dots, p_m^O)$$

with n continuous parameters and m continuous output parameters, where the input parameters must be compared to the condition of the following classifier

$$C(c_1^I, c_2^I, \dots, c_n^I, c_1^O, c_2^O, \dots, c_m^O)$$

with n condition parameters and m action parameters. The continuous (environment) parameters are scaled between 0 and 1, while the condition parameters are determined tertiarily (low:0, high:1, '*don't care*':#) and the action parameters binary (low:0 or high:1).

Two fuzzy sets (*high* and *low*) can now be defined where the membership degree μ_h (high) and μ_l (low) can be attributed for each (scaled) parameter p_i. A representation of this is given in Figure 7.

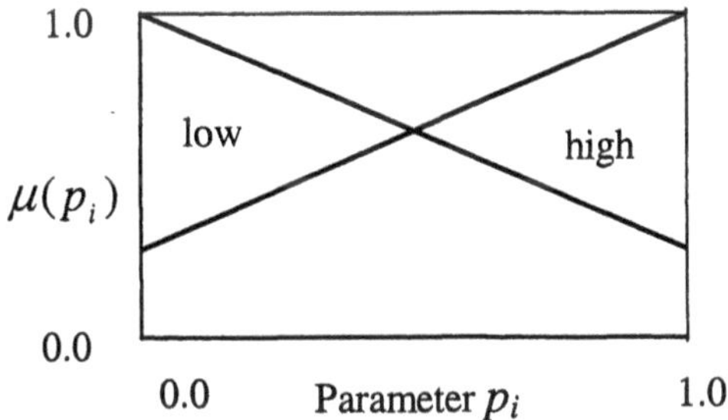

Figure 7: Membership degrees for low and high

The *individual parameter membership degree* $d_i(p_i, c_i)$ of the environment parameter p_i versus the classifier-parameter c_i is defined as:

$$d_i(p_i, c_i) = 1 - \left| p_i^I - c_i^I \right| \text{ when } c_i^I \neq \# \tag{9}$$

$$d_i(p_i, c_i) = 1 \text{ when } c_i^I = \# \tag{9'}$$

in other words the individual membership degree is 1 (100%) when the classifier parameter is a '*#*' or when the environment parameter and the classifier parameter have an identical value (0 or 1). The individual membership degree decreases for values lying apart for the environment parameter and classifier parameter and becomes 0 (0%) when they differ maximally (0 versus 1 or the reverse).

For the input parameters the *total input membership degree* D_I is then determined by:

$$D_I = \prod_{i=1}^{n} d_i(p_i, c_i) \tag{10}$$

An analogous definition can be given to the total output membership degree $D_{O'}$, though without the exception of the symbol $\#$:

$$D_O = \prod_{i=1}^{m} d_i(p_i, c_i) \text{ with } d_i(p_i, c_i) = 1 - \left| p_i^O - c_i^O \right| \tag{11}$$

The value D_I (and D_O) determines the membership degree of the environment message (and action) for each classifier in the ECS. In other words D_I (and D_O) is a measure for the applicability of each classifier concerning the environment message and (action).

Important applications in the ECS algorithm will apply to the definition of the three efficiencies and the BBA selection procedure. The generality $g(t)$ (= the total number of selections) is now determined by D_I and the accuracy $a(t)$ (=total number of successful selections) by D_O where:

$$g(x,t) = g(x,t-1) + D_I(x)$$

and, if eligible for reward:

$$a(x,t) = a(x,t-1) + D_O(x) \text{ for a classifier } x \text{ with } g(x,0) = a(x,0) = 0$$

The definitions mentioned above for g(t) and a(t) are applied for the virtual efficiency $E_v(x,t)$ and real efficiency $E_r(x,t)$.The generality $g(t)$ of the *global* efficiency $E_g(t)$ remains determined by the number of environment messages that are offered to the classifier system, while the accuracy is calculated as indicated above. $E_v(x,t)$ as well as $E_r(x,t)$ can receive a value greater than 1.0 as there can occur selections where $D_O > D_I$, resulting in $a(x,t) > g(x,t)$. As a consequence of the exception for $E_g(t)$ the maximal value for $E_g(t)$ remains limited to 1.0.

The BBA is changed with regard to maximum strength selection, the bid (1.2) and the reward (8):

selection classifier $\quad x = MAX\{D_I(i).S_i(t)\}$

and $\quad P_x(t) = \alpha.D_I(x).S_x(t)$

$$R_x(t) = D_O(x).(\sigma - \frac{E_g(t)}{\varepsilon})$$

In this way the selection of a classifier is based on its strength ***and*** the corresponding input membership degree, while the bid of the classifier is proportional to the input membership degree and the reward proportional to the output membership degree.

4.3 FECS Implementation for the Fiber Yarn Process

In a first application of FECS for the fiber yarn production process the problem is reduced to only continuous input parameters and one discrete output parameter. A fiber yarn process is observed with constant machine settings. The spinnability of the yarn then only depends on the quality of the cotton fibers (20 different fiber qualities were studied).

For each quality the following 5 fiber properties were selected (from the 73 possible fiber characteristics):

fiber length (hviLEN)

uniformity (hviUN)

strength (hviSTR)

elongation (hviEL)

micronaire (hviMIC)

These 5 properties form the continuous (input) parameters of the environment message. The output parameter (or action) is the spinnability of the yarn (0 = non spinnable, 1 = spinnable).

The classifier format was implemented as follows:

LEN	UN	STR	EL	MIC	Spinnability
1 bit	1 bit	1 bit	1 bit	1 bit	1 bit

where the 1-bit coding represents two possible classes: 0 for *low* characteristics and 1 for *high* characteristics. All parameter settings for FECS were kept as determined in Section 3.2, though now with a population of 400 classifiers and

50,000 cycles (1,000 BBA x 50 GA). After the learning phase a final classifier selection was executed in analogy with the procedure described in Section 3.2, though adapted for membership degrees.

A classifier x is selected that meets the condition:

$$x = MAX\{D_I(i).S_i(t)\}$$

AND $E_v(x,t) > \mu.E_g(t)$ for t = 50000 and μ determined by the user (with $0 < \mu \leq 1$)

The generated classifiers are represented in Table 6. Each of these classifiers again has a real physical significance. The first classifier {0###0:1} can be translated by :'if the fiber strength is small (0), as well as the micronaire (0), then the strength (#), elongation (#) and uniformity (#) are unimportant, the spinning machine will be able to spin the yarn (1)'. Contrary to Chapter 3 the reference to a small or big parameter value is now related to a fuzzy set and not to a discrete parameter value.

Table 6: Classifier set for fiber yarn production process (constant machine settings)

LEN	UN	STR	EL	MIC	output	strength	selections	rewards	avg membership degree
0	#	#	#	0	1	1.93	5.98	11	0.54
#	#	1	#	#	0	1.31	2.39	3	0.80
1	#	#	#	#	0	1.29	1.69	2	0.85
0	#	0	#	1	0	0.99	1.95	3	0.65

Using the classifier set above (Table 6) only 1 out of 20 messages is predicted incorrectly (the erroneous message nr. 4 is in Figure 8 indicated with the use of an arrow), resulting in an accuracy of 95%. At the same time environment messages can be classified to their corresponding membership degree (this is the greatest membership degree of the message concerning the classifiers from the classifier set). The result is represented in Figure 9.

Membership degrees vary from 1.0 to 0.26 and are an indication of how well the condition of the message matches the classifier. The chance of a correct prediction depends on the membership degree. A low membership degree indicates a message of which the condition shows little similarity to the resulting classifier set and consequently also has a smaller chance to be predicted correctly by these classifiers. This is illustrated by the erroneous prediction of the fourth message by the classifier {0###0:1} where the corresponding membership degree amount to 0.348 (indicated by Figure 8 using an arrow).

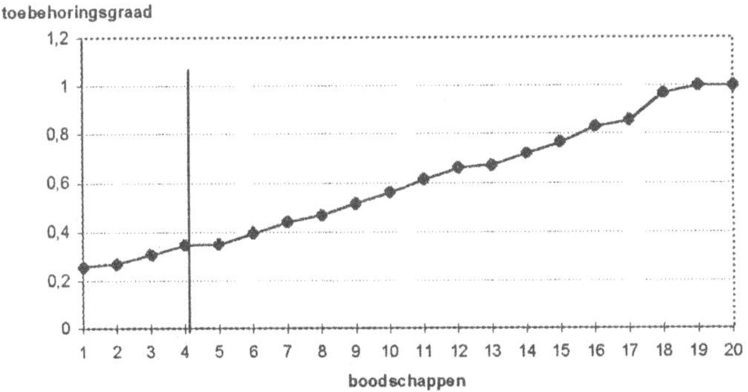

Figure 8: Membership degrees of the messages

For each of the classifiers of Table 6 an average membership degree can be calculated the values of which are represented in the last column of Table 6. It is clear that the first classifier (with the smallest average membership degree of 0.54) has the greatest chance to make an erroneous prediction and effectively caused the only mistake of the classifier set. The three remaining classifiers always have a membership degree greater than 0.6 (= average membership degree of all classifiers) and are faultless.

The importance of the input parameters can be determined as indicated in Section 3.3 and is represented in Table 7

Table 7: Importance of input parameters

	LEN	UN	STR	EL	MIC
Meaningful selections	17	0	6	0	15
Importance	1	-	3	-	2

The most important parameters are fiber length and fiber strength. The fiber uniformity as well as fiber elongation have no effect and so are superfluous/redundant input parameters (within the frame of the chosen constant machine settings).

4.3.1 Combination of Continuous and Discrete Variables in Classifiers

Sections 3 and 4.3 illustrated the operation of ECS (and FECS) for two applications of the fiber yarn process where discrete and continuous parameters respectively were used. As a consequence of these limitations only part of the available data were used. Table 8 gives a short overview of these two applications.

Table 8: Overview of experiments carried out with (F)ECS

	Datasets	Parameters	*Accuracy*
ECS	426	6 (discreet)	94.1%
FECS	20	5 (continuous)	95.0%

The first application (ECS) made use of 4 different fiber qualities and 5 (discretely variable) machine settings, resulting in 6 discrete input parameters. The second application (FECS) is based on 1 (constant) machine setting and 20 different fiber qualities what were determined by 5 continuous fiber characteristics, resulting in 5 continuous input parameters.

The final implementation of FECS will combine the two applications described above: 20 different fiber qualities with 5 different machine settings. The resulting input parameter will consist of 5 continuous input parameters (fiber characteristics) and 5 discrete input parameters (machine settings). The resulting classifier format is consequently

	Discrete					Continuous					
	Yarn count	twist	navel	breaker	rotor	LEN	UN	STR	EL	MIC	
Setting	3	3	2	3	2	2	2	2	2	2	2
Coding	2	2	1	2	1	1	1	1	1	1	1

With a total of 13-bits input coding and 1-bit output coding. The total searching space amount to $3^{13}.21$ = 3,188,646 different classifiers. All data belonging to the 2160 different productions settings (see Section 4.2.2) were uses as learning file for FECS.

FECS is configured as described in Section 4.3. It should be noted that the individual membership degrees (formula 9 and 9') for discrete values remain valid (and consequently also the formulae 10 and 11). For discrete p_i (0 or 1) formula 9 will result in a discrete value (0 or 1). When an individual membership degree is 0, the total input (or output) membership degree ill be equally converted into 0, with as a result that the corresponding classifier is not taken into account for further selection.

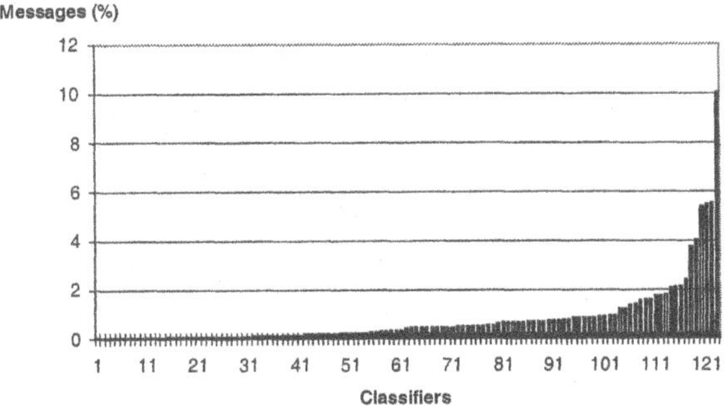

Figure 9: Messages predicted by the generated classifier set

After 993.600 cycles (92 GA x 10,800 BBA cycles) a final classifier set was selected from 123 classifiers resulting in an accuracy of 94.1%. Figure 6.5 shows the importance of each classifier in predicting the environment messages. The classifiers in Figure 4.4 are arranged according to the number of messages they predict. The 6 most important classifiers predict 34.2% of all messages, while the 20 most important classifiers predict 58.1%. The remaining 103 classifiers predict only 36% of the data examples

The input membership degree at the same time gives important information on the applicability of a classifier. Figure 10 gives the % share of the data samples of the input membership degree of the corresponding classifier (this is the classifier that processed the data example). The same Figure also shows the % share of the faulty predictions concerning the input membership degree.

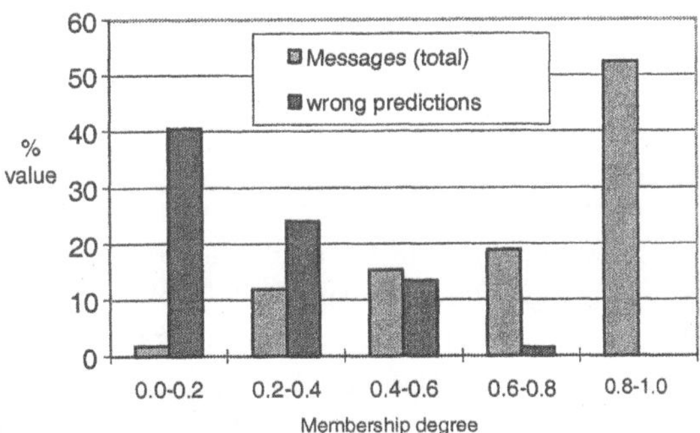

Figure 10: Percent data samples and error distribution according to membership degree

An analysis of Figure 10 leads to the following conclusions:

- the greatest (>50%) part of all data samples (messages)is predicted correctly by classifiers in which the corresponding membership degree is greater than 0.8
- A great (>40%) part of all erroneous predictions occurs by classifiers in which the corresponding membership degree is smaller than 0.2
- when only the classification with a membership degree is considered to be greater than 0.5, than an accuracy of 99% is attained to 80.2% of all data samples.

Table 9: 20 most important classifiers

input	output	fitness	selections	rewards
10100#000####	1	1.1	21.9	26
01000##0#0##0	1	1.8	14.5	26
10#010#0###0#	1	1.1	18.2	29
000#0010##0##	0	1.3	23.6	30
000#0#00##0#1	0	1.6	18.8	33
10010#01#####	1	1.0	34.0	34
01011##1##0##	0	1.3	26.8	34
10100#01#####	1	1.0	37.0	37
10#10#1######	1	1.0	37.0	37
##100#11#####	1	1.0	38.0	38
#0010##100##0	1	2.4	21.1	44
##0##1000#1##	1	2.5	23.1	45
0##01##1###0#	0	1.3	34.9	46
00011##1#####	0	1.0	51.0	51
0#1010#0#####	1	1.0	80.0	80
0#001##1#####	0	1.0	86.0	86
##0110#0#####	1	1.0	115.0	115
01010########	1	1.0	117.0	117
#1100########	1	1.0	118.0	118
####1##1#00##	0	1.7	124.8	214

The 20 classifiers, predicting 58% of all messages, can be considered as the 'most important' classifiers of the generated classifier set. Table 9 gives an overview of these classifiers.The classifier with the highest reward {## ## 1 ## 1 # 0 0 # # : 0} is identical to the most successful (infallible) classifier {0# ## ## 1 ## 1 : 0} generated by ECS in Section 3.3. The first 2 bits (0#), representing the fiber strength are now represented by the (fuzzy) classes STR (and UN) on bit position 11 (0 = low fiber strength) and 10 (0 = low uniformity). Although exactly the same accuracy (94,1%) is attained as in Section 5.4 considerably more classifiers were needed (123 in FECS versus 36 in ECS). This is a direct consequence of the fact that a lot more data samples were dealt with (2160 in FECS versus 426 in

ECS) as well as more input parameters considered (10 for FECS versus 6 for ECS), resulting in more particular cases an a greater diversity of the classifiers.

Figure 11 gives the importance of the input parameters (based on the 20 'most important' classifiers). A parameter with an activation of 80% will only in 20% of all messages dealt with contain a symbol #. The calculation is analogous to the one in Section 3.3.

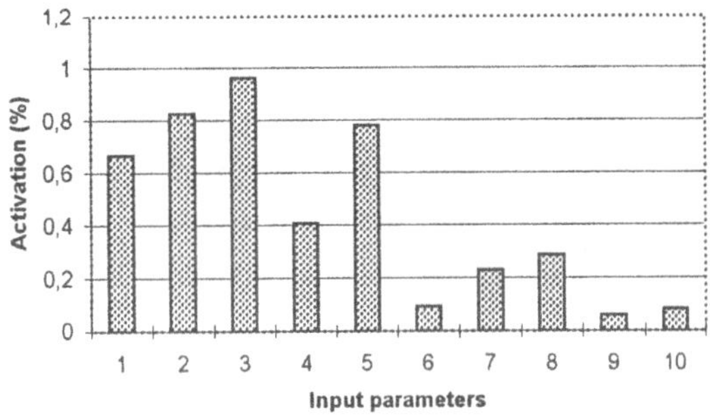

Figure 11: Importance of the input parameters

From Figure 11 it is clear that the machine settings (yarn count, twist, navel, breaker and rotor speed) are generally more important than the fiber quality. None of the fiber characteristics is lacking completely form the generated classifier set. In particular the fiber strength still has an active share in almost 30% of the predicted messages.

4.3.2 Influence of the Number of Fuzzy Sets on the Behavior of FECS

The previous Sections shows the efficiency of FECS for discrete and continuous parameters. All continuous input parameters were implemented using only 2 fuzzy sets. The accuracy of these parameters was therefore limited to an area with a low and a high value, described by the membership degree as illustrated in Figure 7. Applications of fuzzy sets are seldom limited, though, to two classes and can use 3,5 or 7 fuzzy sets in the description of (continuous) input or output parameters. The next Sections will discuss the use and the efficiency of numerous fuzzy sets for FECS.

Suppose a continuous function:

$$f(x_1, x_2, \ldots, x_n) = y \quad \text{met} \qquad f : \Re^n \rightarrow \Re$$

and a corresponding classifier-format:

$$c(x_1, x_2, \ldots, x_n, y)$$

where each parameter (input and output) is coded in k fuzzy sets, resulting in a searching space of $k.(k+1)^n$ possible classifiers (the condition of the classifier has $k+1$ possible values including the *'don't care'* symbol). In the least favorable case (*'worst case function'*) there is no generalization (no # symbols) possible for each classifier. Each classifier describes consequently a unique input-output -function value. This exhaustive image of the function results in k^n classifiers. It is clear that k should be chosen as small as possible to reduce the dimensions of the searching space and also decrease the number of possible different classifiers (that describe the function). If k is chosen too small the final accuracy of the classifier set will be very limited.

The following (simple) function $f(x) = x$ illustrates this by comparing the two different codings namely the coding in 2 fuzzy sets ($k = 2$) and 7 fuzzy sets ($k = 7$). The corresponding search space and the number of generated classifiers is mentioned in Table 10.

Table 10: Search space and classifiers for $f(x) = x$ and $k = 2$, $k = 7$

	2 vaagverzamelingen	7 vaagverzamelingen
Zoekruimte	6 mogelijke classifiers	56 mogelijke classifiers
generated classifiers	2 (00 en 11)	7 (00,11,22, ... , 66)

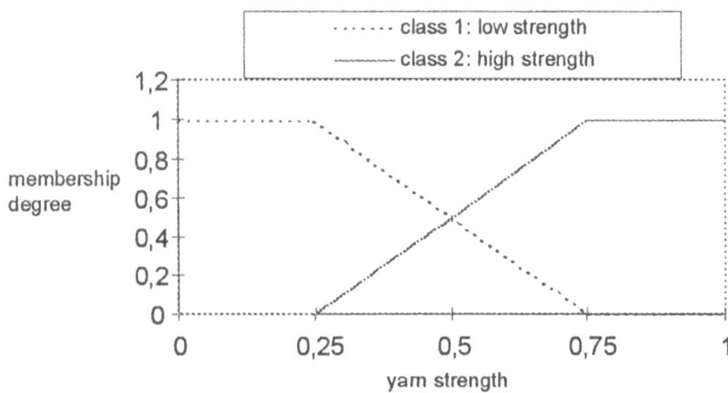

Figure 12: 2 fuzzy classes for yarn strength

As appears from the example above the search space and the number of generated classifiers strongly depends on the number of considered fuzzy sets. The classifier description for $k = 2$ is very limited though: for a low (high) $x = 0$ (1) value a low (high) output = 0 (1) is achieved. Many functions (different from the identical function) will meet this description. This is not the case for $k = 7$, giving a clearer development of the function $f(x) = x$.

The application of $k = 7$ on the continuous parameters of the fiber-yarn production process as described in section 6.3.5 would create a search space with $3^8.3^{15}.2 \approx 188.10^9$ classifiers. The search space is almost 60,000 times bigger that the originally (in Section 6.3.5) studied search space ($k = 2$). As already 123 classifiers were generated in this configuration ($k = 2$), numerous thousands can be expected for $k = 7$. Although this is theoretically possible this last configuration (k= 7) exceeds the possibilities (and processor capacities) of the presently implemented model.

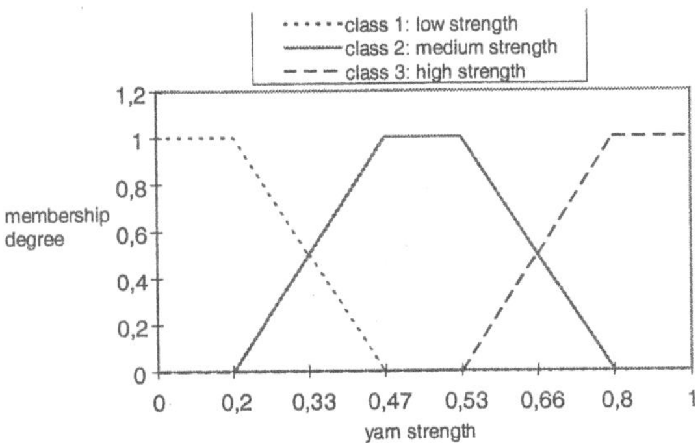

Figure 13: 3 fuzzy classes for yarn strength

Although the implementation of numerous (more tan 2) fuzzy sets for all input and output parameters is practically impossible, this is feasible for one (or a few) important parameters without increasing the search and classifier space considerably. Fro the selected parameter a higher accuracy is achieved in this away. Previous studies have applied the FECS model on the fiber-yarn production process aiming at the prediction of the spinnability of the yarn. The following application will study the possibility to predict yarn strength based on the implementation of numerous (2,3 and 5) fuzzy sets.

4.3.3 Implementation of Fuzzy Sets of Type k=2, k=3 and k=5 for the Prediction of Yarn Strength

An analogous classifier format and FECS configuration as in Section 6.3.5 is used for this: the input consists of 5 fuzzy (2classes) fiber properties and 5 discrete machine settings, resulting in a 13-bit coded condition. The output (action) of the classifier will consist of the (continuous) yarn strength. The learning file is now limited to the 1380 spinnable production settings where a strength of the produced yarn could be determined. The non spinnable yarns are indeed not eligible as no final yarn strength can be attributed. The coding of the yarn strength happened respectively in 2, 3 end 5 fuzzy classes. The degrees of belonging for each of these classes is given in the Figure 12 (k =2), Figure 13 (k = 3) and Figure 14 (k = 5).

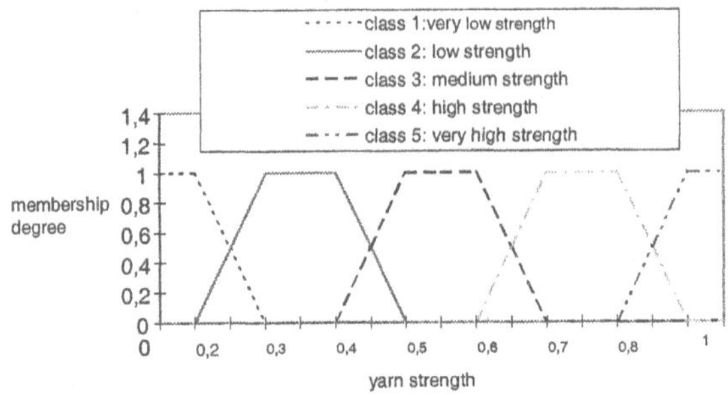

Figure 14: 5 fuzzy classes for yarn strength

The yarn strength is rescaled in the Figures above to a value between 0 (lowest yarn strength) and 1 (highest yarn strength). The degrees of belonging of the respective classes are chosen so that each describes an equal part of the yarn strength (the surface under each of the curves is the same of each class). A histogram of the data samples of the learning file concerning the yarn strength is given in Figure 14 and shows an uneven strength distribution.

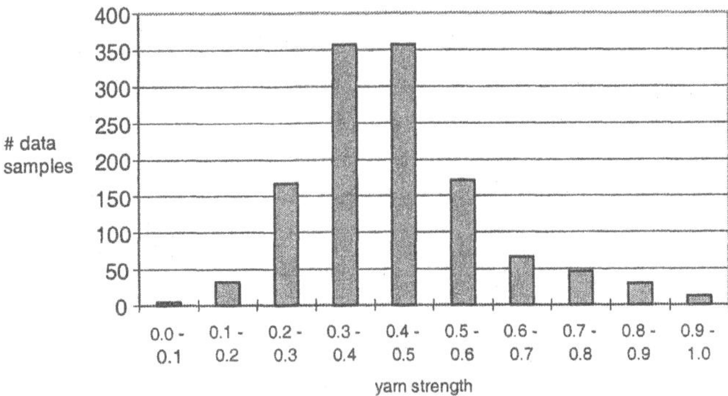

Figure 15: Yarn strength histogram

The membership degrees as indicated in Figures 13 and 14 could lead to classes with few data samples (as e.g. class 1 and 5 in Figure 14), while the 'middle'-class (as class 3 in Figure 14) would contain a proportionally huge number of data samples. This can lead easily to the submission or total elimination of one of the extreme (low or high) classes during the learning process. To avoid this problem 'new' data examples with a low and high yarn strength can be created artificially by adding in the learning file randomly chosen copies of existing data samples with a low/high strength. A number of yarn examples with medium strength were also eliminated from the learning file. The resulting learning file consist of 650 data samples distributed evenly over the yarn strength (65 examples for each 0.1 increment in yarn strength).

Table 11 gives the number of learning cycles, selected population growth and the results for the $k = 2$, $k = 3$ and $k = 5$.

Table 11: Results for k=2, 3 and 5

k	Cycles	size	Accuracy	error < 1 class	error = 1 class	Nr of classifiers
2	100k	200	93.5%	6.5%	0%	44
3	200k	400	92.0%	8.0%	0%	62
5	1M	500	80.3%	10.9%	8.8%	119

The achieved classifier sets have 2, 3 or 5 fuzzy (strength) classes as possible output (action). When the (original) continuous strength is distributed in 2,3 and 5 discrete classes (the border situated on the co-ordinates for which the membership degree is 0.5) the accuracy of the achieved classifier set can be calculated by observing the procentual number of data samples that falls in a (correct) corresponding discrete class. When this is not the case two possibilities can occur:

- the considered data example has a membership degree concerning the correct class that is smaller than 0.5, though greater than 0.0. The procentual share of these errors is represented in column 5 of Table 11 (error < 1 class)
- the considered data example has a membership degree of 0.0 concerning the correct class. The procentual share of these error is represented in the sixth column of Table 11 (error>= 1 class).

The last column gives the number of classifiers in the finally selected classifier set (for the corresponding configuration of k).

Using Table 11 the following conclusions can be reached:

- To attain an acceptable convergence (and accuracy)the number of learning cycles must be increased drastically when k increases (a factor 10 for k=2 to k=5). An analogous conclusion is valid for the necessary classifier population (a factor 25 for k=2 to k=5). The total learning time for k=5 then is 25 times greater than for k=2
- Accuracy decreases with increasing k, resulting for k =5 in 8.8% error (error = 1 class). *This does not at all indicate to a somewhat worse result, as the classes-resolution also increases for higher k-values.*

- The number of generated classifier needed to predict the presented data samples, increases for higher *k* values (from which follows also the requirements of a higher classifier population).

The error distribution for k=2 is represented in Figure 16.

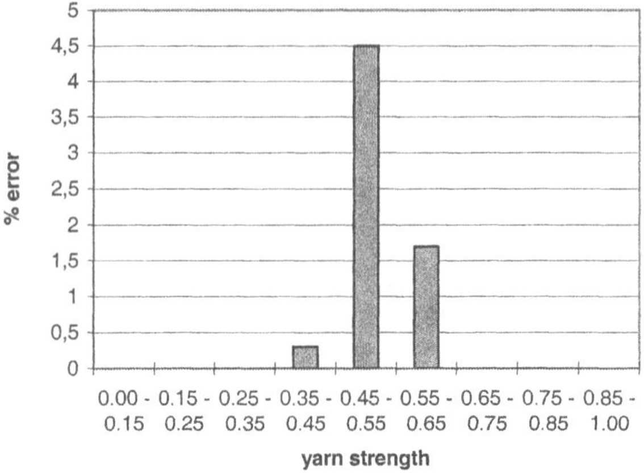

Figure 16: Error distribution for k=2

Table 12: Classifiers for *k*=2

condition	output	strength	selections	Rewards
#1##100##11##	1	1.1	13.0	13.0
010#001###1##	1	1.0	14.5	14.2
1#01##0#0##1#	1	1.3	14.2	18.4
1#0####1#1###	1	1.1	22.0	22.7
010#0#01#1###	1	1.1	23.8	23.8
1##0####0#0##	0	1.1	25.6	24.8
0#0#0##0#1###	1	1.1	28.4	28.5
1###10###1###	1	1.1	31.3	31.2
1#0#00####1##	1	1.3	36.6	45.0
0##0####0#0##	0	1.6	147.6	201.6

Although no 'real' mistakes were made (membership degree for all 'errors' is also way smaller than 0.5 though higher then 0.0 for the correct class), the membership degree indicates a preference for another (incorrect) class. From Figure 16 it is clear that all these error (for *k*=2) are situated in the transfer zone (medium strength) between the classes with high strength and with a membership degree between 0.2 and 0.8. The classification in 2 classes can be considered as 100% acceptable an thus (very) successful. For *k*=3 almost similar results are reached.

The only difference is found in the membership degree of the 'dubious' data samples (8%) which now varies from 0.0 to 1.0. This indicates a higher error (in comparison to *k*=2) though there are still no 'real' errors (>= 1 class) generated. For *k*=5 8.8% 'real' errors are generated distributed over the entire strength area. This indicates a loss in accuracy in distributing into 5 fuzzy sets.

Table 13: Classifiers for *k*=3

Condition	Output	strength	selections	rewards
1##0#000#11##	2	1.1	14.3	14.8
010100###11##	2	1.1	15.6	15.4
0100#000##1##	2	1.1	16.8	16.8
10###0#1###0#	1	1.1	16.1	16.9
#1#010#00#0##	0	1.4	16.0	17.9
1#000#####1##	2	1.2	19.2	20.5
0#0000##0#00#	0	1.3	16.9	21.9
0#000011 0#0##	0	1.4	18.9	23.8
10#1######1##	2	1.2	41.3	48.2
000#10#00#0##	0	1.4	90.3	117.2

Tables 12, 13 and 14 give an outline of the 10 most important classifiers generated for *k*=2, *k*=3 and *k*=5.

Some of the classifier characteristics are summarized in Table 15.

With:

- The min# (second column) the minimum number of #-symbols that occur in a classifier
- The max# (third column) the maximum number of #-symbols occurring in a classifier
- the #/classifier (fourth column) the average number of # symbols for a classifier
- % predictions (fifth column) the number of data samples that were classified by the 10 most important classifiers

This leads to the following remarks and conclusions:

For increasing k the average number # symbols for a classifier decreases. As a result the classifiers for *k*=5 are a lot less generalized that those for *k*=2 and *k*=3 In fact, there is hardly any space (#symbols) left for *k*=5 for further classifier specification. This indicates that for *k*=5 a limit is approached where for higher *k* there will be no generalization in the classifiers, resulting in a useless classifier set (in which only copies of the decreasing number of # symbols is observed in the %

number of predictions by the classifier set: k=2 predicts more than 75% of all examples, while k=5 predicts only 26%.

Table 14: Classifiers for k=5

Condition	Output	strength	selections	rewards
10010#00#11##	4	1.4	7.1	9
00000011##0#0	0	1.6	7.7	10.3
01001000111##	4	1.1	11.0	10.4
10001000111##	4	1.1	10.2	12.0
00001000##0#0	1	2.1	8.2	13.6
1001#0#1#11##	4	1.3	11.8	14.1
10000001#0##0	2	1.9	8.5	14.7
000010000#001	0	2.2	10.2	19.0
000010100#0#0	0	2.3	9.2	19.8
000010100#001	0	2.6	20.3	48.1

When the generated classifiers for k=2 and k=3 are compared than the two most important classifiers of both classifiers sets appear to show similarities. These classifiers predict the lowest (0) as well as the highest (1 or 2) strength.

Fiber strength (position 11) plays an important part in predicting the yarn strength (which is intuitively acceptable). Other fiber characteristics are less important in comparison to the machine settings. These tendency is also observed for k=5, where most of the (remaining) #symbols are situated in the fiber characteristics.

Table 15. Classifier characteristics for k=2, 3 and 5

K	min #	max #	# / classifier	% predictions
2	6	9	7.8	75.9%
3	4	9	6.1	60.0%
5	1	5	2.3	26.3%

For k=3 and k=5 there are hardly any classifiers within the set of the 10 most important classifiers (1 classifier for k=3 and 2 classifiers for k=5) predicting a medium yarn strength. This is a result of the equalization of the histogram where a greater variation of data samples occurs for the medium yarn strength, resulting in a greater diversification of classifiers. So it is dangerous (in this example) to limit the classifier set to the 10 most important classifiers. Final predictions must take account of all generated classifiers.

The final stage would consist of predicting again a continuous yarn strength starting from the generated classifier sets. For this use could be made of a defuzzification procedure of the attained result generated by the classifier set. The defined membership degree degrees of the 3 (and 5) fuzzy sets as given in Figure 13 (and 14) are unsuitable for defuzzification as the membership degree degrees have a trapezoidal form. A consequence of this is the existence of three (or five) areas in which the membership degree degrees is always 1.0 and there is no possibility to reach a further distinguishing in such an area. For k=3 this is in the areas [0.0,0.2] (low strength), [0.47, 0.53] (medium strength) and [0.8, 1.0] (high strength). Within these intervals there is no variation of membership degree degrees, so there can be attained no more accurate continuous value for the parameter (in the intervals mentioned above). A 'triangular' form for the membership degree degrees would male a more efficient defuzzification possible, though this would also result in an uneven number of data samples for each class or a variation in the form for each class.

Without modification of the FECS-configuration of Section 4.3 (and the results following from that) a simple (approaching) defuzzification can be reached by:

a selection of a classifiers set where

the strength is higher than 0.5

$E_v(x_i) \geq \mu . E_g$ met $\mu = 0.8$

a continuous yarn strength y calculated by:

$$y = \frac{\sum_{i=1}^{k} D_I(x_i).c(x_i)}{\sum_{i=1}^{k} D_I(x_i)} \tag{12}$$

with $c(x_i)$ the output of the corresponding classifier x_i

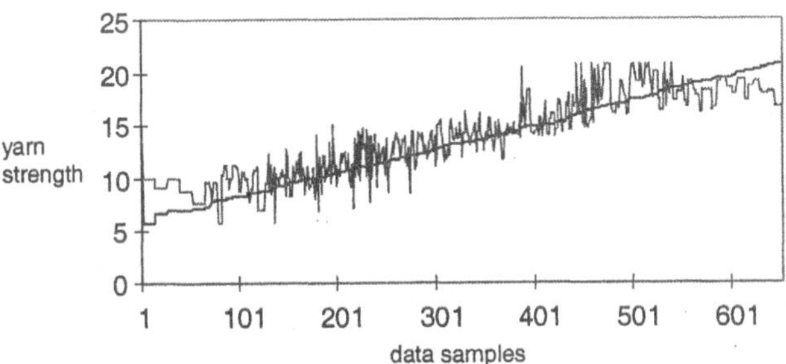

Figure 17: Continuous yarn strength for *k*=3

The result for *k*=3 and *k*=5 is represented in Figures 17 and 18.

On the continuous curve in Figures 17 and 18 the experimental (measured) yarn strength values are classified according to increasing strength. The second (irregular) curve are the predicted (defuzzified) values. For *k*=3 a procentual accuracy of 86.6% is reached where great deviations occur in the lowest and highest part of the yarn strength spectrum. These deviations are a consequence of the information loss by the specific formation of the membership degree degrees. A better result is reached for k=5 where the procentual accuracy amounts to 90.9%. Although the classification in 5 classes originally seemed less suitable (see Table 11) the final result is considerably better than for k=3. The errors made in k=5 (in Table 11) are the consequence of the higher resolution attained in the distribution of the strength spectrum to 5 fuzzy sets. These errors do not at all indicate a less accurate model in comparison to k=3. The attained accuracy for k=5 differs only 3.4% with the results attained using the backpropagation model (Sette, 1997).

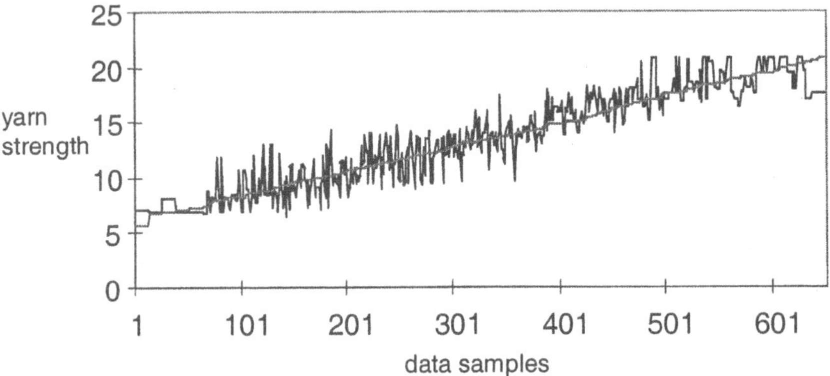

Figure 18: Continuous yarn strength for k=5

5 Conclusions

On the basis of this chapter the following conclusions can be formulated:

The ECS-configuration is successfully extended by introducing to fuzzy classes for input and output parameters so that the possibility now exists to process continuous coded surrounding messages. In order to do this the BBA and the classifier efficiencies were redefined using (total) input and output membership levels. The resulting classifier system, named FECS successfully generated a classifier set for some simple continuous functions. FECS was also applied to the fiber yarn production process (to predict spinnability) where a classifier set was generated predicting with 94.1% accuracy the spinnability of the yarn based on the (continuous) fiber characteristics and the machine settings. Besides the input membership level of appears to be a guideline of a classifier concerning the environment message.

It was also demonstrated that FECS allows processing a continuous parameter using more than two fuzzy classes. This must be limited to one (or some) important parameter(s) needing a higher precision as otherwise the observed search space an the number of generated classifiers soon becomes great. The prediction of the yarn strength in the fiber yarn production process using 3 fuzzy classes lead to an accuracy of 92% based on the defined classes, making not a single substantial error (error of class 1). The prediction of a continuous yarn strength, by applying a defuzzification to 5 classes, finally resulted in a procentual accuracy of 90.9%.

6 References

Bonnarini A. (1997a), Anytime learning and adaptation of structured fuzzy behaviors, accepted for publication on the Special Issue of the Adaptive Behavior Journal about "Complete agent learning in complex environments", Maja Mataric (Ed.), No. 5

Boullart L., Sette S. (1997), Genetic Algorithms: Theory & Applications, Journal A, Vol. 38, n.2, 13-23.

Boullart L., Sette S. (1998), High Performance Learning Classifier Systems, Engineering of Intelligent Systems (EIS'98), International ICSC Symposium, 11-13 February, Tenerife, 249-256.

BRITE/EURAM project BREU 00052-TT (1990-1993), 'Research for a mathematical and rule based system which allows to optimize a cotton mixture, based on the interdependence of significant fiber properties, process parameters, yarn properties and spinning machinery performances.'

Darwin C. (1859), The Origin of Species: by Means of Natural Selection or the Preservation of Favored Races in the struggle for Life.

Goldberg D.E. (1989), Genetic Algorithms in Search, Optimization and Machine Learning., Addison Wesley Publishing Company.

Holland J.H. (1968), Hierarchical description of universal spaces and adaptive systems, Tech. Report ORA Projects 01252 and 0826, Ann Arbor, University of Michigan, Dept. Comp. Science & Comm. Science

Holland J.H. (1973), Genetic Algorithms and the Optimal Allocation of Trials., SIAM Journal of Computing, Vol. 2 (2), 88-105

Holland J.H., Reitman J.S. (1978), Cognitive systems based on adaptive algorithms. In D.A. Waterman & F. Hayes-Roth (Eds.), Pattern directed inference systems (pp.313-329). New York: Academic Press

Pynckels F., Sette S., Van Langenhove L., Kiekens P., Impe K. (1995), Use of Neural nets for Determining the Spinnability of Fibers, Journal of the Textile Institute, 86 (3), 425-437

Pynckels F., Kiekens P., Sette S., L. Van Langenhove, Impe K.. Use of Neural Nets to Simulate the Spinning Process, The Journal of the Textile Institute, 88 (1), 440-447

Sette S., Boullart L., Van Langenhove L., Kiekens P. (1997a), Optimizing the Fiber-to-Yarn Production Process with a combined Neural Network/Genetic Algorithm Approach, Textile Research Journal, Vol. 67, No. 2, 84-92

Sette S., Learning systems by means of evolutionary algorithms, PhD thesis, Faculty of Applied Sciences, University Ghent, 1998.

List of Contributors

Charles Bock
School of Science and Health
Philadelphia University,
Philadelphia, PA 19144, USA

E.L. Brannon
Department of Consumer Affairs
Auburn University
Auburn, AL 36849 USA

Shu-Cherng Fang
Department of Industrial Engineering
North Carolina State University
Raleigh, NC 27695-7906, USA

Ashish Garg
School of Textiles and Materials Technology
Philadelphia University,
Philadelphia, PA 19144, USA

P. Kedziora
Institute of Mechanics & Machine Design,
Cracow University of Technology
ul. Warszawska
Kraków, Poland

Russell E. King
Department of Industrial Engineering
North Carolina State University
Raleigh, NC 27695-7906, USA

T. Marshall
Department of Consumer Affairs
Auburn University
Auburn, AL 36849 USA

A. Muc
Institute of Mechanics & Machine Design,
Cracow University of Technology
ul. Warszawska
Kraków, Poland

Henry L.W. Nuttle
Department of Industrial Engineering
North Carolina State University
Raleigh, NC 27695-7906, USA

Christopher M. Pastore.
Philadelphia University
School of Textiles and Materials Technology
Philadelphia, PA 19144, USA

Stefan Sette
Department of Textiles
University of Gent
Gent, Belgium

Les M. Sztandera
Philadelphia University
Computer Science Department
Philadelphia, PA 19144, U.S.A.

S. Thommesen
Department of Consumer Affairs
Auburn University
Auburn, AL 36849 USA

Mendel Trachtman
School of Science and Health
Philadelphia University,
Philadelphia, PA 19144, USA

Lieva Van Langenhove
Department of Textiles
University of Gent
Gent, Belgium

Janardhan Velga
Philadelphia University,
Philadelphia, PA 19144, USA

James R. Wilson
Department of Industrial Engineering
North Carolina State University
Raleigh, NC 27695-7906, USA

Bugao Xu
Department of Human Ecology
The University of Texas at Austin
Austin, TX 78712, USA

About the Editors

Dr. Les M. Sztandera is an Associate Professor of Computer Science and Head of the Computer Science Program at the Philadelphia University, Philadelphia, Pennsylvania, U.S.A. Prof. Sztandera has been named Distinguished Fulbright FLAD Chair in Information Systems at the Technical University of Lisbon, Portugal, for the 2002-2003 academic year.

He has been involved in soft computing teaching and research since 1987. Dr. Sztandera has 12 years of full time university teaching experience, and is a recipient of a Teaching Excellence Award. He developed a sequence of soft computing courses coupled with laboratory assignments in which students work with real life problems, such as detecting an industrial pollutant, predicting strength and density of materials, designing a medical expert system, simulating protective systems in complex power generating units, detecting carcinogenic dyes, or designing new drugs.

Complementary with his teaching effort, Dr. Sztandera has been involved in a variety of research activities. That has resulted in numerous research grants from the Department of Commerce, National Textile Center, National Science Foundation, Ohio Supercomputer Center, Pittsburgh Supercomputer Center, and American Heart Association. Over $1,000,000 in research funding has been experienced. Those research activities also resulted in 30 journal publications and 50 conference presentations.

Dr. Sztandera received his Ph.D. degree from the Department of Electrical Engineering and Computer Science, University of Toledo, Ohio, U.S.A., with a dissertation on Fuzzy Sets in Self-Generating Neural Network Architectures. He earned his M.Sc. degree from the Department of Computer Science and Engineering, University of Missouri, Missouri, U.S.A., with a thesis on Spatial Relations Among Fuzzy Subsets of an Image, and a Diploma in English from University of Cambridge, Cambridge, England.

Dr. Sztandera is a member of professional organizations in the U.S. and Canada: the North American Fuzzy Information Processing Society, Association for Computing Machinery, and Canadian Society for Fuzzy Information and Neural Systems. His scientific and scholarly research contributions to the fuzzy set theory are internationally recognized. He proposed, designed, and implemented fuzzy neural trees. For this and other contributions to the fuzzy sets and systems theory, he was included in the Encyclopedia of Computer Science and Technology, 1999 Edition. Dr. Sztandera is also listed in the Marquis Who's Who in the World, Who's Who in Science and Engineering, Who's Who in America, and Who's Who in the East.

Dr. Christopher Pastore is an Associate Professor of Textile Engineering, Head of Textile Engineering program, and Director of Research for the School of Textiles and Materials Technology at Philadelphia University in Philadelphia, Pennsylvania, USA. Prior to his arrival at Philadelphia University, he was an Associate Professor of Textile Materials Science at North Carolina State University, College of Textiles, Raleigh, NC, USA, and before that was Assistant Professor of Materials Engineering and Associate Director of the Fibrous Materials Research Center at Drexel University in Philadelphia.

Dr. Pastore is the author of several books and over 50 publications related to the properties and applications of textiles and their composites and has presented his work world-wide. His research focuses on the structure-property relationships of textiles which has led to interest in the area of soft-computing.

Dr. Pastore has a Ph.D. in Materials Engineering from Drexel University, Philadelphia, PA, USA, a M.S. in Mathematics from Drexel University, and a B.A. in Mathematics from LaSalle University, Philadelphia, PA, USA. He is the recipient of the 2001 Fiber Society Award for Distinguished Research in Basic or Applied Fiber Science, and has received a commendation from the City of Philadelphia for his work with high school educators and students in the field of applied engineering science.

He is quite active in research, having participated in funded programs sponsored by agencies such as NASA, US-EPA, US Navy, US Air Force, St. Gobain, Ford Motor Company, Chrysler Corporation, Atlantic Research Corporation, Northrop-Grumman, and others.

www.ingramcontent.com/pod-product-compliance
Ingram Content Group UK Ltd.
Pitfield, Milton Keynes, MK11 3LW, UK
UKHW021826190726
13853UKWH00003B/1219

* 9 7 8 3 6 6 2 0 0 2 9 2 6 *